SERICULTURE

Cultivation, Processing and Marketing

The Author

Dr. Ramakrishna Naika completed his Graduation, Post Graduation and PhD in Sericulture from University of Agricultural Sciences, Bangalore. He received Dr. B.R. Ambedkar Gold Medal for securing highest CGPA at Masters Degree level. He Received ICAR Best Teacher Award for his contribution and excellence in teaching. Dr. Ramakrishna Naika has handled four externally funded projects sponsored by Department of Science and Technology, University Grants Commission and Rastriya Krishi Vikas Yojana, Government of India. He has participated in the national and international conferences and seminars, presented papers and chaired the technical sessions. Further, he served as member of board of examiners and member of expert committee for selection of teachers under promotion in University of Mysore and SKUAST, Jammu. He has guided 10 M.Sc (Seri.) and 01 PhD student as Chairman and served as member of Advisory Committees for 12 Post Graduate students. Published more than 100 scientific papers in National and International Journals apart from publishing 60 popular articles in reputed magazines. He has also authored/edited twelve books. Presently he is serving as Professor and Head, Department of Crop protection and Sericulture at College of Sericulture, Chintamani.

Dr. P. Venkataravana, obtained his M.Sc. (Agri.) in Genetics and Plant Breeding and Ph.D. from UAS, Bangalore and guided 15 M.Sc. and 02 Ph.D. students. He published 150 research papers in reputed National and International Journals and published 2 books, 05 books chapter and released 02 high yielding varieties of groundnut, 01 high yielding variety of tamarind and Jamun. For his excellence in applied research he received Nagamma Dattathreya Rao Desai award from University of Agriculture Sciences, Bangalore. Presently he is working as Dean (Seri.), College of Sericulture, Chintamani, University of Agricultural Science, GKVK, Bangalore.

Dr. Sunil Kumar T., obtained his M.Sc. (Agri.) in Sericulture and Doctor of Philosophy in Sericulture from University of Agricultural Sciences, Bangalore. Awarded two gold medal for Ph.D degree programme and qualified ICAR NET. He published 18 research papers in reputed National and International Journals, he published two books in Kannada literature and received state level prize from Karnataka Science Academy and Karnataka Saahitya Academy for his excellence in Kannada literature. He has participated as resource person to deliver talk on Sericulture in DD-Chandana TV programme and All India Radio programme. Presently he is working as an Assistant Professor (Contractual) in Department of Sericulture, College of Sericulture, Chintamani, University of Agricultural Sciences, GKVK, Bengaluru.

SERICULTURE

Cultivation, Processing and Marketing

— Author —

Ramakrishna Naika

P. Venkataravana

Sunil Kumar T.

2024

Daya Publishing House®

A Division of

Astral International Pvt. Ltd.

New Delhi – 110 002

ISBN 978-93-5461-967-0

Publisher's Note:

Every possible effort has been made to ensure that the information contained in this book is accurate at the time of going to press and the publisher and author cannot accept responsibility for any errors or omissions, however caused. No responsibility for loss or damage occasioned to any person acting, or refraining from action, as a result of the material in this publication can be accepted by the editor, the publisher or the author. The Publisher is not associated with any product or vendor mentioned in the book. The contents of this work are intended to further general scientific research, understanding and discussion only. Readers should consult with a specialist where appropriate.

Every effort has been made to trace the owners of copyright material used in this book, if any. The author and the publisher will be grateful for any omission brought to their notice for acknowledgement in the future editions of the book.

Published by : **Daya Publishing House®**
A Division of
Astral International Pvt. Ltd.
– ISO 9001:2015 Certified Company –
4736/23, Ansari Road, Darya Ganj
New Delhi-110 002
Ph. 011-43549197, 23278134
E-mail: info@astralint.com
Website: www.astralint.com

Printed at : **Replika Press Pvt Ltd**

PREFACE

Silk is the protein secreted by the larvae of certain Lepidopteran moths. It is fibrous in nature, and widely used for the manufacture of art clothes. The silk is derived from silk moths mainly belonging to Bombycidae and Saturuidae families of lepidoptera. The origin of silk industry is a controversial topic. According to one source the origin is from India during 200 B.C. (Himalayan region). Whereas another view holds that it originated from China. The knowledge of silk production spread far and wide by the passage of time. Today many Western nations along with Japan, Russia, Persia, Africa and India are practicing the culture and production of silk. Sericulture is the only industry that gives a large number of job opportunities. So, the process of Sericulture is considered a tool for the economic reconstruction of rural areas. Sericulture has been identified as one of the important components for economic development in India. Its uniqueness in the fact that the sericulture activities not only engage the rural households in the cultivation of mulberry and silkworm rearing but also encompass in their fold a whole range of reelers and weavers.

In nature, Asia is the major manufacturer of silk in the world and produces in excess of 95 per cent of the entirety global output. India is the leading manufacturer of silk and also have largest consumer of silk in the world. India has a well-built tradition and culture bound home market of silk. In India, mulberry silk is produced mainly in the states of Karnataka, Andhra Pradesh, Tamil Nadu, Jammu and Kashmir and West Bengal. The non-mulberry silks are produced in the state like Jharkhand, Chhattisgarh, Orissa and north-eastern states. The first silk filament cloth was made in Ancient China. Silk filament is a fine, touch, elastic fibre which is produced from caterpillar's named 'bombyx mori'. Within two or three days spinning it surrounds itself in a cocoon which consists of a continuous and very fine filament of silk.

Then silkworm turns into a moth, which escapes by making a hole through the cocoon. The first cocoons were placed in very hot water to soften the sticky gum, sericin, which holds the filament together. The ends of the filament are found and it would be unwound. This progression is called reeling. The silk filament is put on to large reels which are known as swifts. Silk is not like cotton or wool; it is not spun but twisted. Throwing includes the revolving of two sets of bobbins at different, carefully adjusted, speeds.

The present book entitled *Sericulture: Processing, Cultivation and Marketing* includes different aspects of Sericulture spread over in nineteen chapters, which covers introduction, moriculture, mulberry cultivation, pests and diseases of mulberry, propagation methods in mulberry, advanced micropropagation methods, irrigation, manuring and pruning, harvesting and storage of mulberry, botany of bombyx mori, morphology of bombyx mori, anatomy of bombyx mori, seed technology, by-products of sericulture, silk reeling, silkworm extension, Govt. schemes and subsidy, sericulture tools, appliances and machines, value addition to sericulture, silkworm marketing and economics and climatic factors for higher silk production. The text is substantiated with many figures and illustrations that will help better understanding of the text. The book, which is primarily intended for the undergraduate students of agriculture, will be of use to the postgraduate students of entomology and sericulture, fellow teachers, farmers, naturalists, policy makers and to the officials working in the department of agriculture, especially those involved in sericulture and also to the elite public who are interested in sericulture.

We are grateful to all those persons as well as various books, manuals, periodicals, magazines, journals *etc.* that helped in the preparation of this book. In spite of the best efforts, it is possible that some errors may have occurred into the compilation and editing of the book. Further queries, constructive suggestions and criticisms for the improvement of the book are always welcome and shall be thankfully acknowledged.

CONTENTS

Chapter 1

Introduction: An Overview

1.0. Introduction

Silk is one of the oldest fibres known to man. It is an animal fibre produced by certain insects to build their cocoons and webs. Although many insects produce silk, only the filament produced by the mulberry silk moth *Bombyx mori* L. and a few others in the same genus is used by the commercial silk industry. The silk produced by other insects, mainly spiders, is used in a small number of other commercial applications, for example weapon and telescope cross-hairs and other optical instruments.

Over the centuries, silk has been regarded as a highly valued textile fibre. Its qualities of strength, elasticity, softness, absorbency, affinity for dyes and adaptability to various forms of twisting continue to meet various applications. Because of its

The Indian sericulture market was worth INR 307.4 Billion in 2020. Sericulture refers to the art of rearing silkworms for the production of silk. India is currently the world's second largest producer of raw silk and the largest consumer of raw silk and silk fabrics. The market for silk in India is driven by both exports and a very strong domestic demand. Fabrics made up of silk are quite popular in the domestic market during ceremonies, religious rituals, weddings, festivals, etc. Silk is used as a raw material for the manufacturing of both garments such as suites, sarees, etc. as well as in products such as curtains, bed sheets, pillow covers etc. Although silk is currently regarded as a luxury item in India with its price being significantly higher than other fabrics, we expect a continuous increase in disposable incomes to increase the consumption of silk fabrics in the country.

high (tensile) strength, lustre, durability and ability to bind chemical dyes, silk is still considered a leading textile material. Despite facing keen competition from man-made fibres, silk has maintained its supremacy in the production of luxury apparel and other high-quality goods. Silk fibres display unusual mechanical properties: they are strong, extensible and mechanically compressible. Silk is rightly called the 'queen of textiles' for its lustre and feel. Silk's natural beauty and properties of comfort in warm weather and warmth during colder months have also made it ideal for high-fashion clothing. As a result there is significant research into the artificial production of silk fibres.

Sericulture is the art of rearing silkworms for the production of cocoons which are the raw material for the production of silk. The major activities of sericulture comprise food-plant cultivation to feed the silkworms which spin silk cocoons and reeling the cocoons to unwind the silk fi lament for value added benefits such as processing and weaving. Sericulture is ideally suited for improving the rural economy as it is practised as a subsidiary industry to agriculture. Recent research has also shown that sericulture can be developed as a highly rewarding agro-industry. Sericulture involves the cultivation of mulberry and production of cocoons to produce silk filaments. The best raw silk is obtained from the species of moth *B. mori*. Breeding of silkworm normally occurs once in a year but, under industrial conditions, eggs may be hatched three times a year. The female moth lays around 350–400 eggs, after which the moths die. As they are subject to hereditary infection, any eggs from infected moths are destroyed. Larvae of about 3 mm are hatched from the eggs. For about 20–30 days, they are carefully nurtured and are fed five times a day on chopped mulberry leaves. In the meantime, the larvae change their skin four times and are formed into caterpillars about 9 cm long. At this point they are ready to spin a cocoon, for which racks, clusters of twigs or straw are provided.

The caterpillars have small openings under their jaws called spinnerets, through which they secrete a protein-like substance. This substance solidifies when it comes in contact with air and the resulting fi lament is spun around the silkworm in a shape resembling the digit 8. The cocoon, which is about the size of a peanut shell, is completed. The fi lament is held together by sericin or silk gum. The life of the worm is ended by the process of 'stoving' or 'stifling' in which the cocoons are heated. Some of the cocoons are preserved so that the pupae or chrysalises inside them develop into moths for further breeding.

There are five major types of silk of commercial importance, obtained from different species of silkworms which in turn feed on a number of food plants. The main type is mulberry. Other varieties of silks are generally termed non-mulberry silks. India has the unique distinction of producing all these commercial varieties of silk.

1.1 Sericulture

Sericulture is an agro-based industry. It involves rearing of silkworms for the production of raw silk, which is the yarn obtained out of cocoons spun by certain species of insects. The major activities of sericulture comprise of food-plant cultivation to feed the silkworms which spin silk cocoons and reeling the cocoons for unwinding the silk filament for value added benefits such as processing and weaving.

Although there are several commercial species of silkworms, *Bombyx mori* is the most widely used and intensively studied. According to Confucian texts, the discovery of silk production by *B. mori* dates to about 2700 BC, although archaeological records point to silk cultivation as early as the Yangshao period (5000 – 10,000 BCE). By the first half of the 1st century AD it had reached ancient Khotan, and by AD 140 the practice had been established in India. Later it was introduced to Europe, the Mediterranean and other Asiatic countries. Sericulture has become one of the most important cottage industries in a number of countries like China, Japan, India, Korea, Brazil, Russia, Italy and France. Today, China and India are the two main producers, together manufacturing more 90 per cent the world production each year.

1.1.1 Sericulturist

Innovative researchers have turned sericulture into an industry and now it has become a major cash crop of the country. The sericulture industry is unique for more than one reasons. It is based on agricultural output specifically cocoons and cottage-based labor intensive in nature.

1.2 History, Distribution and Production

1.2.1 History

There is no authentic information regarding the origin and use of silk. The ancient literature gives two views. According to one view, silk industry originated for the first time in India at the foot of the Himalayas, and from there it spread to other countries of the world. Second view, which has greater acceptance, says that this industry originated in China about 2700 B.C. According to this, a Chinese Princess Xi Ling Shi was the first to discover the art of reeling an unbroken filament from a cocoon. This art was kept a close secret for nearly 3000 years. This art later on spread to the rest of the world through several agencies like civil war refugees, war prisoners' marriage of royal families *etc.*

1.2.2 Distribution of Silk

Silk is classed as an animal fibre, since it is derived from the cocoon of a moth, but it has little in common with wool. In being a continuous filament, it has more in common with synthetic textiles. It is the finest and therefore one of the most beautiful, of natural fibres and is thus highly-prized.

The difficulties of dealing with so fine a filament as that produced by the silkworm make silk an expensive, luxury fabric. The art of making textiles from silk was discovered in China, around 2,700 B.C. and the process was kept secret and monopolized by the Chinese until it gradually spread to Japan and India in the third and fourth centuries A.D. Knowledge of the process only reached Europe in the sixth century A.D.

The special qualities of silk which make it so sought after are its fineness and its glossy appearance when made up. Its properties cannot yet be successfully matched by other natural or synthetic textiles, but its production and use has declined because of its high cost and labour-intensive methods.

1.2.3 World Production of Silk

Total raw silk production today is relatively small, amounting to only about 47,000 tonnes annually, compared with 14 million tonnes of cotton, and almost all of it is produced in Asia. The leading producers are Japan (34 per cent), China (34 per cent), South Korea, the U.S.S.R., and India. The only non- Asian producer of any significance is Brazil. France, Britain, Italy, Switzerland and the U.S.A. are silk-manufacturing countries, though their output of raw silk is infinitesimal.

World production of silk has tended to decline during the twentieth century for two reasons. Firstly, the costs of production are high and have been increasing in most areas. This means that silk is an expensive fabric and its market is limited. Secondly, the development of synthetic textiles, especially rayon, which in continuous filament form resembles silk so much that it is called artificial silk, has created great competition.

Artificial silk is cheap and easy to produce, unlike genuine silk, and is therefore used for all non-luxury purposes. Natural silk, however, retains its luxury market and in the last two decades has even increased its market among the wealthy members of the U.S., European and Japanese society.

Japan

Silk has been produced in Japan for about 1,600 years, since the art was first introduced from China. The rearing of silkworms is a traditional occupation of the rural population in central and southern Honshu, especially in upland areas where other agricultural possibilities were few and where silk production supplemented farm income. Peak production was attained before the Second World War when about 2.25 million people were engaged in silk production but the war and post-war period saw a decline in output. Since 1950, however, the industry has again expanded but on a different basis. In pre-war days raw silk was exported and in 1929 accounted for about 40 per cent of Japan's total exports. Nowadays silk is only a small and relatively unimportant feature of the Japanese economy because so many other industries have been rapidly developed. Moreover, silk is no longer exported raw but is mostly manufactured in Japan. The number of people engaged in silk production is now also much smaller. Rural population has drifted to the towns and only 10 or 15 per cent of farms now produce silk. The acreage under mulberry trees has remained stagnant. Silk production, however, has been greatly increased by the improvements of rearing techniques and the breeding of higher-yielding silkworms. Costs of production are no longer low because the standard of living has risen, but the Japanese are able to compete with other producers because of their greater efficiency. Japan is the largest producer of silk textiles as well as of raw silk.

China

China was the first country to produce silk, nearly 4,500 years ago, and has always been a major producer. The most important areas of production are in central and southern China and in the Shandong (Shantung) peninsula, and the silk is produced largely by peasants as a supplement to farm income. Both raw and manufactured silk are exported. China is a competitive producer because it has large supplies of labour and a long tradition of production. Silk production was fairly important in Europe in the eighteenth and nineteenth centuries but the rise in the standard of living and the increase in alternative employment opportunities in both France and Italy have reduced local production. Silk manufacture, originally based on locally-produced silk is now dependent on imported supplies. France, Italy, and Switzerland, which also has a tradition of silk manufacturing, are major silk importers. Lyons and Grenoble in France and Milan in Italy are the main centres. Silk is also manufactured in Krefeld, Germany and at Macclesfield in Britain. The largest silk importer, however, is the U.S.A. Both raw silk, woven in New England, and silk textiles are imported to supply the affluent American market. European and some minor producers, such as Thailand, depend on the tourist trade to supply a market for their silk goods.

1.3 Silk

Man is always inquisitive for silk products. SILK - The Queen of Textiles, spells luxury, elegance, class and comfort. Mankind has always loved this shimmering fibre of unparalleled grandeur from the moment Chinese Empress Shiling Ti discovered it in her tea cup. It withstood many a daunting challenge from other natural and artificial fibres and yet, remained the undisputed Queen of Textiles since centuries. Exquisite qualities like the natural sheen, inherent affinity for dyes and vibrant colours, high absorbance, light weight, resilience and excellent drape *etc.* have made silk, the irresistible and inevitable companion of the eve, all over the world. Chemically speaking, silk is made of proteins secreted in the fluid state by a caterpillar, popularly known as 'silkworm'. These silkworms feed on the selected food plants and spin cocoons as a 'protective shell' to perpetuate the life. Silkworm has four stages in its life cycle *viz.*, egg, caterpillar, pupa and moth. Man interferes this life cycle at the cocoon stage to obtain the silk, a continuous filament of commercial importance, used in weaving of the dream fabric.

1.3.1 Habit, Habitat and Life History

Out of the four different silk types the two *i.e.*, mulberry and Eri are manufactured from domesticated silkworms, whereas Tasar and Munga silkworms are wild in nature, although attempts are in progress to domesticate them too. The life-cycle of these four types of silk moths are much in common, as they lay eggs, from which caterpillars hatches. They eat, grow and produces cocoon for their protection, then pupate inside cocoon. After sometime moths emerge from the cocoon, male and female mate, lay eggs, and repeat their lifecycles. The characteristic feature of these silk-producing moths is that they spin a cocoon of silk for the protection of their pupae. The man with his mental superiority has discovered the technique of robbing the silk threads from these cocoons for his own use.

1.3.2 Types of Silk

There are four major types of silk of commercial importance, obtained from different species of silkworms which in turn feed on a number of food plants. Moths belonging to families Satumidae and Bombycidae of order lepidoptera and class Insecta produce silk of commerce. There are many species of silk-moth which can produce the silk of commerce, but only few have been exploited by man for the purpose. Mainly four types of silk have been recognized which are secreted by different species of silk worms.

- ☆ **Mulberry Silk:** This silk is supposed to be superior in quality to the other types due to its shining and creamy white colour. It is secreted by the caterpillar of *Bombyx mori* which feeds on mulberry leaves.
- ☆ **Tasar Silk:** It is secreted by caterpillars of *Antheraea mylitta, A. paphia, A. royeli, A. pernyi, A. proyeli etc.* This silk is of coppery colour. They feed on the leaves of Arjun, Asan, Sal, Oak and various other secondary food plants.
- ☆ **Eri Silk:** It is produced by caterpillars of *Samia cynthia ricini* which feed on castor leaves. Its colour is also creamy white like mulberry silk, but is less shining than the latter.
- ☆ **Munga Silk:** It is obtained from caterpillars of *Antheraea assama* which feeds on Som, Champa and Moyankuri.

1.3.4 Silk Production in India

There are five major types of silk of commercial importance, obtained from different species of silkworms. These are Mulberry, Oak Tasar and Tropical Tasar, Muga and Eri. Except for mulberry, other non-mulberry varieties of silks are wild silks, known as vanya silks. India has the unique distinction of producing all these commercial varieties of silk. South India is the leading silk producing area of the country and is also known for its famous silk weaving enclaves like Kancheepuram, Dharmavaram, Arni, *etc.*

1.3.5 Silk Industry in India

Silk is a lustrous, tough, elastic fibre produced by the larvae of silkworms; the term also covers the thread or cloth made from this fibre. The silk industry originated in China, as early as 2700 BC according to tradition. Towards the 3rd century AD, knowledge of the silkworm and its product reached Japan through

The Present Status of Indian Sericulture

Particulars	*Unit*	*2019-20*	*2020-21*	*2021-22*
Mulberry Plantation	ha.	239967	237578	242277
Mulberry Cocoon Production Total	**MT**	185143	169345	187241
Mulberry Silk Production				
Total	**MT**	**25239**	**23896**	**25818**
Vanya Silk				
Tasar Reeling Cocoon	Lakh No.	24181	20775	12217
Tasar Silk Yarn	MT	3136	2689	1466
Eri Cut Cocoon	MT	9219	8951	9528
Eri Spun Yarn	MT	7204	6946	7364
Muga Reeling Cocoon	Lakh No.	12814	12056	12886
Muga Raw Silk	MT	241	239	255
Total Vanya Silk Production	**MT**	**10581**	**9874**	**9085**
Total Raw Silk Production	**MT**	**35820**	**33770**	**34903**
Exports	**Crore Rs.**	**1745.65**	**1466.6**	**1848.96**
Import of Raw Silk				
Quantity	MT	3315	1804	1978
Value	Crore Rs.	1149.32	570.56	819.68
Employment	Lakh Person	94.30	87.3	87.8

State-wise Raw Silk Production in India (Unit ;MT)

Sl. No.	*State*	*2019-20*		*2020-21*		*2021-22*	
		Target	*Achmnt.*	*Target*	*Achmnt.*	*Target*	*Achmnt. (P)*
1	Karnataka	12000	11143	12600	11292	12500	11191
2	Andhra Pradesh	7946	7962	8208	8422	9305	8835
3	Telangana	295	297	310	309	337	404
4	Tamil Nadu	2300	2154	2300	1834	2400	2373
5	Kerala	20	13	17	7	10	9
6	Maharashtra	630	428	475	428	560	523
7	Uttar Pradesh	365	309	354	316	395	354
8	Madhya Pradesh	165	61	80	47	74	30
9	Chhattisgarh	562	480	535	300	561	224
10	West Bengal	2900	2295	2520	872	1630	1632

Sl. No.	State	2019-20		2020-21		2021-22	
		Target	Achmnt.	Target	Achmnt.	Target	Achmnt. (P)
11	Bihar	86	56	58	64	96	46
12	Jharkhand	2604	2402	2904	2185	2902	1046
13	Odisha	150	137	160	102	185	108
14	Jammu and Kashmir	170	117	142	80	150	98
15	Himachal Pradesh	50	31	45	20	40	28
16	Uttarakhand	42	40	25	25	42	42
17	Haryana	2	1	1	1	1	1
18	Punjab	5	3	4.5	1	2	4
19	Assam	5395	5316	5519	5462	5855	5700
20	Arunachal Pradesh	75	64	67	43	59	53
21	Manipur	600	504	542	327	530	502
22	Meghalaya	1220	1192	1245	1213	1367	1234
23	Mizoram	130	104	113	43	59	59
24	Nagaland	682	600	649	264	311	315
25	Sikkim	1	1	2	0.08	5	0.03
26	Tripura	130	111	125	112	125	113
	Total	**38530**	**35820**	**39000**	**33770**	**39500**	**34923**

Korea; it probably spread to India a little later. India is the largest consumer of silk in the world. Silk is a unique fabric which has gradually crept in everybody's home in some way or the other. The Indian silk industry is an integral part of the Indian textile industry and is among the oldest industries in India. It engages around 60 lakh workers and it involves small and marginal farmers. Indian silk industry is the second highest contribution of silk to the world production. Consistent market demand and remarkable approach has led to the widening of this industry. India exports considerable quantity of silk, goods and the value of these silk goods is more than 15 per cent of the total raw silk production. India s export includes Dress Materials, Readymade Garments, Saris and made-up articles for interior decoration. The Indian silk goods are being exported to the traditional major markets like the USA and European countries and small markets of Asia Region.

☆ India is the second largest producer of silk in the world. It provides employment to over 87.8 Lakh people in the country.

- ☆ There are four major types of silk produced in India: Mulberry, Tasar, Muga, Eri of which Mulberry accounts for 70 per cent of total raw silk production.
- ☆ India currently produces all four variety of silk - Mulberry, Eri, Muga and Tasar. The silk production is mostly prevalent in Karnataka, Assam, West Bengal, Tamil Nadu, Andhra Pradesh and Jammu and Kashmir.
- ☆ Major Export destination of Indian Silk exports are USA and UAE followed by UK, France, Italy and Germany. Mostly natural silk yarns, fabrics, made-ups, readymade garments, silk carpets and silk waste are exported.

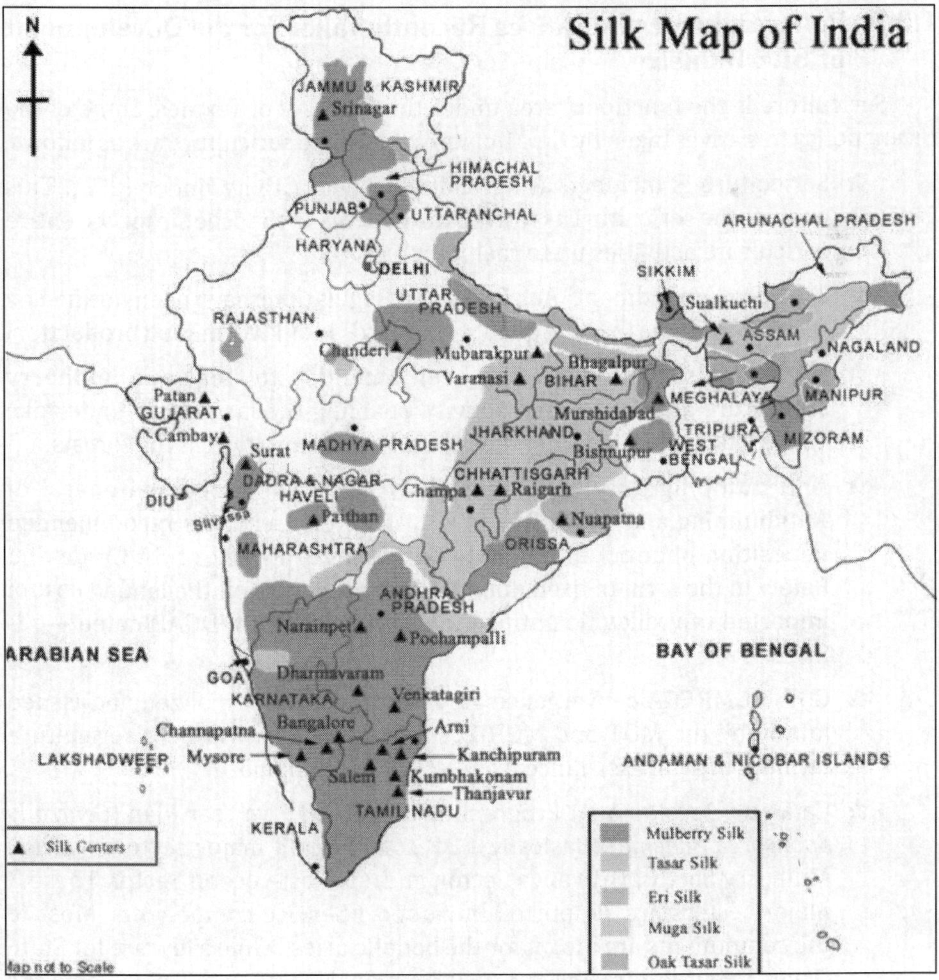

Silk Map of India
***Source*: Google (Smart Edge).**

- For growth and development of the silk industry Indian Silk Export Promotion Council has also been set up. It organizes tradeshows and fairs across the world to promote trade with different countries. The council also facilitates meetings between exporters and potential customers.
- Handlooms account for about 85 per cent of silk consumption in India while power looms use the remaining.
- India's north eastern region has the unique distinction of producing all these commercial varieties of silk contributes about 21 per cent of the total silk production in the country.

1.3.6 Policy/Schemes Initiatives Recently Taken for the Development of Silk Industry

Sericulture is the functional area under the Ministry of Textiles. Some of the recent policy initiatives taken by the Ministry to promote sericulture are as follows.

- Sericulture is included as agriculture allied activity under RKVY. This enables the sericulturists to avail the benefits of the scheme for the entire sericulture activities up to reeling.
- The CSB (Amendment) Act, Rules and Regulations have been notified by the Govt. of India to bring quality standards in silkworm seed production.
- Forest Conservation Act has been amended to treat non mulberry sericulture as forest-based activity enabling the farmers to undertake Vanya silkworm rearing in the natural host plantation in the forests.
- Anti dumping duty on Chinese raw silk - The Director General of Antidumping and Allied Duties (DGAD), New Delhi has recommended imposition of antidumping duty on Chinese raw silk of 3A Grade and Below in the form of fixed duty of US$ 1.85 per Kg on the landed cost of imported raw silk vide notification No.14/17/2014/DGAD dated 4-12-2015.
- CDP-MGMREGA convergence guideline have been finalized and issued jointly by the MOT and MORD. These guidelines will help sericulture farmers to avail assistance from MGNREGA scheme.
- Catalytic Development Programme during XII Five Year Plan (*Centrally Sponsored Scheme*): Basically, CDP comprises 3 major sectors, *viz.* (a) Mulberry Sector; (b) Vanya Sector, and, (c) Post-cocoon Sector. These 3 major sectors are supported by Support Service components. Most of the components are meant for the beneficiaries, while a few are for State Govts. Towards infrastructure development at State level. The scheme has been approved for implementation with a total project cost of Rs.1,852.36 crore, of which the GoI share is Rs.889.00 crore, State share is Rs.426.87 crore and balance Rs.536.49 crore is beneficiary share.

- ☆ The Scheme: "SILK SAMAGRA" has been approved by the Government of India for continuation beyond XII Plan for three years from 2017-18 to 2019-20 at a total allocation of Rs.2161.68 crore with the following four components: 1. Research and Development, Training, Transfer of Technology and I.T. Initiatives, 2. Seed Organizations, 3. Coordination and Market Development and 4. Quality Certification Systems (QCS)/Export Brand Promotion and Technology Upgradation.
- ☆ Multi-State Project for Promotion of Large Scale Tasar Sericulture Based Livelihoods under Mahila Kisan Sashaktikaran Pariyojana (MKSP)
- ☆ North East Region (NER) Textile Promotion Scheme (Sericulture Projects for NE during XII Plan), In order to boost the textile sector in the NE region, the Govt of India has approved a project- based strategy for the North East Region under an umbrella scheme by name "North East Region Textile Promotion Scheme". This umbrella scheme will be implemented in the North East Region in addition to the regular schemes of the Ministry. (Source: Ministry of Textiles, Government of India)

1.3.7 Employment Generation

The employment generation in the country is 8.8 million persons (provisional) in 2021-22 compared to 8.7 million persons in 2020-21, indicating increase of 1.1 per cent.

Sericulture Industry in India has classified the employment generation pattern of the industry into two major types:

- ☆ ***Direct Employment:*** Mulberry Cultivation; Leaf Harvesting and Silk Worm Rearing.
- ☆ ***Indirect Employment:*** Reeling; Twisting; Weaving; Printing and Dyeing; Finishing and Silk Waste Processing.

1.3.8 Significance of Sericulture Industry

Sericulture is an agro-based and economically rewarding enterprise consisting of several sets of activities and plays a predominant role in shaping the economy destiny of the rural people with lot of employment potentially.

The production process of silk fabric contains a long chain of interdependent operations *i.e,* cultivation of mulberry, silkworm seed production, silkworm rearing, silk production, twisting, warp and weft making, dyeing and printing, spun silk production, finishing of silk fabric, designing of silk garments *etc.*, that provides medium of live hood to rural and semi urban people.

India is blessed with favourable climatic conditions throughout the year and availability of human resource remained the advantage to the Indian silk industry.

Sericulture provides rich dividends with low investment and portable return within short gestation period and provides employment throughout the year. it is obvious that, Sericulture plays a pivotal role in economic development of the country by generating employment, income, as well as foreign exchange.

1.3.9 Problems of Sericulture Industry

Sericulture is ideally suited for improving the rural economy of the country, as it is practiced as a subsidiary industry to agriculture, it is hindered by various factors like imports of cheap and alternative textiles from other Asian neighbors, use of outdated manufacturing technology, primitive and unscientific "reeling" and "weaving" techniques, use of poor quality seeds, low production of bivoltine seeds, use of non-graded and diseased seeds, poor knowledge of farm disease amongst farmers, poor supply chain management, huge unorganized and decentralized sector, high production cost, recurring droughts and increased import of silk from China and accompanied with the common problems faced by the Indian silk industry are – Price fluctuation, Absence of proper market, Lack of transport facilities, Absence of storage facilities, Poor information on market trend, Lack of finance, High cost of production and low productivity, lack of technology penetration, prevalent rural nature of the industry *etc.*, competition from synthetic fibers.

Low export earnings – due to global recession and reduced demand in western countries for silk goods. A weaker rupee is also hurting exports. However, the silk exports are finding non-traditional/new markets in UAE, Nigeria, Thailand *etc.*

Selling power loom in the name of handloom – leads to inadequate returns on hard work of handloom workers since power loom is much cheaper.

Declining inclination of youth towards weaving – as one can earn the same money working at a power loom with less stress.

Competitive pricing - The blending of cheap imported Chinese silk or artificial/ synthetic silk yarns putting the natural silk trader son the verge of distress sales.

Decline in area cultivated - Mulberry silk in the country has seen a steady decline in its area of mulberry cultivation because of rapid urbanization, industrialization and a shortage of agricultural labour.

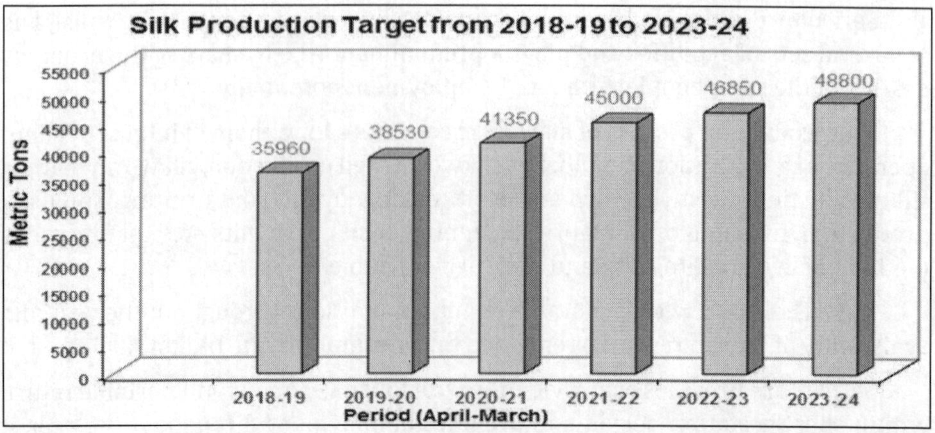

Piece meal approach of government in term of banning foreign silk, lack of integrated market and inadequate knowledge of sericulture amongst the traders. Because of all these factor India could not reached the target level to become global leader at international silk market

1.3.10 India Silk Industry at a Glance 2021-22

India is the second largest producer of silk in the world. Sericulture or the silk industry or the rearing of silk worms provided employment to over 87.8 Lakh people in India during 2022. Sericulture activities in India are spread across 52,360 villages. India is known for its commercially useful varieties of silk.

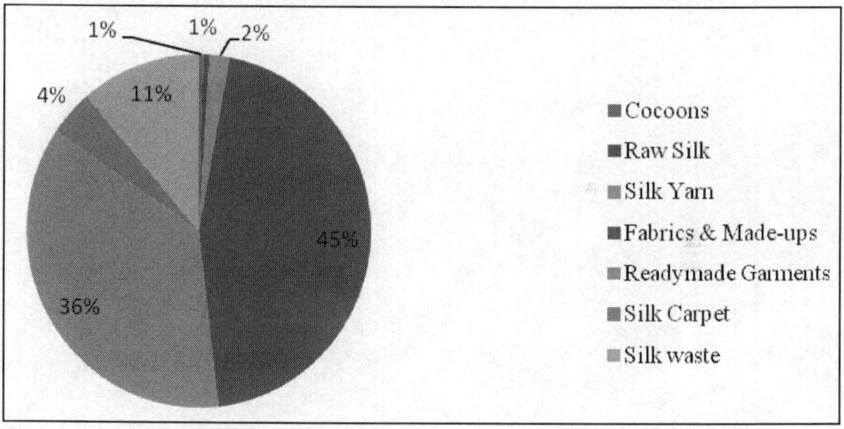

Silk Export Earnings from Silk and Silk Goods
(*Source*: Central Silk Board. Bengaluru).

Chapter 2

Moriculture

2.1 Introduction to Moriculture

Moriculture can be defined as the process of cultivating mulberry plants in order to obtain feed for silkworms. In India, about 3,00,000 hectares of farmland is known to be committed to the cultivation of mulberry plants. Initially, the approximate yield of each hectare was known to range from 12000 to 15000 kilograms of mulberry leaves. However, advancements in the field of agricultural technology have brought about a great increase in the productivity of these farmlands.

For example, the plantation of high yielding varieties of the mulberry plant and the proper maintenance of the mulberry plantation can result in each hectare offering a yield ranging from 25000 to 35000 kilograms of mulberry leaves. It is important to note that moriculture plays a vital role in the practice of sericulture.

Area under Mulberry in Different States

State	*Area(ha)*
Andhra Pradesh	38084
Assam	2813
Jammu and Kashmir	4717
Karnataka	166000
Kerala	1164
Madhya Pradesh	2043
Manipur	*25975

State	Area(ha)
Tamil Nadu	9491
Uttar Pradesh	5665
West Bengal	21358
Other	4934

2.2 Geographical Distribution of Mulberry Plant

The utilization of and research on mulberry in different parts of the world are depicted as under:

Country Summary of Mulberry Area, Use and Research

Country		Utilization				Research		
	Area	Silk	Fruit	Forage	Other*	Agr.	Breed.	Feed
AFRICA								
Egypt and Tunisia	n.a.**	Ö	Ö			Ö		
Ethiopia	n.a.	Ö				Ö		
Kenya	n.a.	Ö		Ö		Ö		Ö
Madagascar	n.a.	Ö				Ö		
Tanzania	n.a.			Ö				Ö
AMERICAS								
Argentina and Bolivia	n.a.				Ö1			
Brazil	38	Ö		Ö		Ö	Ö	Ö
Colombia	n.a.	Ö		Ö		Ö		
Costa Rica	n.a.			Ö		Ö		Ö
Cuba	<1			Ö		Ö		Ö
Dominican Rep.	<1			Ö				
El Salvador	<1			Ö				Ö
Guatemala	n.a.			Ö				Ö
Honduras and Panama	<1			Ö				
Mexico	<1	Ö	Ö	Ö	Ö1	Ö		Ö
Panama	<1			Ö				
Peru	n.a.				Ö1			
Saint Vincent	<1			Ö				
United States	n.a.				Ö1			Ö

Country		Utilization				Research		
	Area	Silk	Fruit	Forage	Other*	Agr.	Breed.	Feed
ASIA								
Afghanistan	n.a.			Ö				
China	626	Ö			Ö2	Ö	Ö	
India	280	Ö		Ö	Ö3	Ö	Ö	Ö
Indonesia	n.a.	Ö						
Japan	n.a.	Ö	Ö	Ö	Ö2	Ö	Ö	Ö
Korea	n.a.	Ö		Ö		Ö	Ö	
Kyrgyzstan	n.a.	Ö	Ö			Ö		
Malaysia	n.a.	Ö				Ö		
Pakistan	n.a.	Ö		Ö		Ö		
Philippines	n.a.	Ö				Ö		
Syrian Arab Republic and Turkey	n.a.		Ö					
Tajikistan	n.a.	Ö				Ö		
Turkmenistan	n.a.	Ö	Ö			Ö		
Viet Nam	n.a.	Ö		Ö		Ö	Ö	
Uzbekistan	n.a.	Ö				Ö		
EUROPE								
Bulgaria	n.a.					Ö		
France	n.a.			Ö	Ö1	Ö		Ö
Greece	n.a.				Ö1			
Italy	n.a.				Ö1	Ö		Ö
Poland	n.a.					Ö		
Spain	n.a.				Ö2			

* Other uses: 1: Landscaping and gardening; 2: Medicinal and infusion; 3: Handicrafts and cabinet work

** n.a. data not available.

2.3 Classification of Nursery

Kingdom	Plantae-Plant
Subkingdom	Tracheobionta-Vascular plant
Superdivision	Spermatophyta
Division	Magnoliophyta-Flowering plant
Class	Magnoliopsida-Dicotyledons

Subclass	Hamamelididae
Order	Urticales
Family	Moraceae
Genus	*Morus* L
Species	*Morus alba, Morusaustralis, Morus cathayana, Morus macroura, Morus mongolica, Morus nigra, Morus notabilis, Morus serrate, Morus celtidifolia, Morus insignis, Morus microphylla, Morus rubra, Morus mesozygia* and *Morus indica*

2.4 Ideal Climatic Conditions for Moriculture

It can be noted that the climatic conditions required for moriculture can be found in almost all states in India. The ideal temperature for the growth of mulberry plants is known to be 24 degrees Celsius. However, these plants can thrive in the temperature range of 18 to 38 degrees Celsius.

Due to the relatively high irrigation requirements of the mulberry plant, the ideal season for the plantation of mulberry plants is the pre-monsoon season.

2.4.1 Favourable Soil for Moriculture

The type of soil plays a significant role in the optimum and healthy growth of any plant. Despite the fact that the mulberry plant can grow in a wide range of soils, it does have certain soil requirements. For example, since the mulberry plant is a deep-rooted plant, it requires sandy loamy/red sandy loamy soil which has a good capacity to hold water.

This plant also thrives in relatively porous soils in which the air can penetrate to relatively deep regions. Finally, the plant is known to thrive in the soils that are slightly acidic in nature. The ideal pH range of the soil for moriculture ranges from 6.5 to 7.5.

2.4.2 Variety

Mulberry is the sole food plant of silkworm, *Bombyx mori.* More than 300 species belongs to the genus *Morus;* which includes *Morus alba, Morus nigra, Morus indica, Morus laevigata, Morus multicaulis* and others however *Morus alba* and *Morus indica* are generally found in tropical and sub- tropical regions.

In Karnataka, different varieties of mulberry include Mysore Local, Kanva-2 (M5). S-54, S-30, S-41, S-36, G2 and G4 are being cultivated by the sericulturists under irrigated condition. Mysore local, M5, S-13 (for red soil), S-34(for black cotton soil), RFS-135, RFS-175 are the varieties recommended for rain fed areas. However S-36, V-1 and G2 are recommended for chawki silkworms.

2.4.3 Propagation of Mulberry Plants in Moriculture

It can be noted that the mulberry plant can be propagated via two different methods, namely sexual mulberry propagation and asexual mulberry propagation. The sexual propagation of the mulberry plant involves the plantation of seeds. Before the plantation of seeds, the land must be suitably prepared and the seeds must be filtered based on their quality. Care must be taken to maintain at least 6 per cent moisture in the seeds during their storage.

The asexual propagation of the mulberry plant can be achieved via one of three different methods, as listed below.

- ☆ Propagation of mulberry plant via cutting
- ☆ Propagation of mulberry plant via grafting
- ☆ Propagation of mulberry plant via budding

It is important to note that the most common method that is used for the asexual propagation of the mulberry plant is through mulberry cuttings.

2.4.4 Planting System

Pit system of planting with wider spacing in rainfed mulberry should be adopted. Pits of size of 35x35x35cm are dug at 90x90cm spacing. Pits are to be filled with 2kg FYM. However in irrigated condition Pit or trench system with wider spacing and row system with wider or closer spacing is followed.

Different spacing's are followed in mulberry plantation (1' x 1' to 5' x 5'), of late, paired row system of planting with (90 x 60cm) x 150cm and for tree type of mulberry 10 m X 10 m also being adopted for aiding mechanization in cultivation

2.5 Leaves Harvesting from the Mulberry Plant

Once the mulberry plants have grown to a certain extent, their leaves can be harvested and used as feed for silkworms. The harvesting of the mulberry leaf is a time-sensitive task since fully mature mulberry leaves have relatively lower nutritional value than the leaves of the partially mature mulberry plant.

If the leaves are not harvested at the right time, the lack in the nutritional value of the mulberry leaves will manifest as diseases in the silkworm that consume these leaves. The two most common methods of harvesting leaves from the mulberry plant are leaf picking and branch cutting. Another important method for the harvesting of mulberry leaves is the harvesting of the entire shoot (also known as whole shoot harvesting).

Component	*Percentage (per cent)*	*Component*	*Percentage (per cent)*
Moisture	65-78	Protein	19-25
Minerals	10-15	Reducing sugar	1.2-1.9
Sugars	10-15		

2.5.1 Harvesting of Mulberry Leaves via Leaf Picking

The harvesting of mulberry leaves via leaf picking involves the plucking of individual leaves from the mulberry plant. Therefore, this method of leaf harvesting in moriculture is considered to be highly labour-intensive.

It is interesting to note that this is the most common method that is employed for the harvesting of mulberry leaves in India. It can also be noted that mulberry leaves can be harvested approximately five to six times from each plant in a year. The types of leaves that are harvested via leaf picking usually depends on the stage the silkworm has reached in its life cycle. For example, relatively young silkworms are usually fed with tender leaves whereas the relatively old silkworms are fed slightly mature mulberry leaves.

2.5.2 Harvesting of Mulberry Leaves via Shoot Cutting

This method of harvesting mulberry leaves involves the cutting of entire Shoots. The advantages of this method of mulberry leaf harvesting over the harvesting of mulberry leaves via leaf picking are listed below:

- Since it is easier to cut off the entire Shoots than picking off each individual leaf, the branch cutting method of mulberry leaf harvesting is less labour intensive. Therefore, the labour costs associated with this method is also lower.
- The lifespans of the leaves are relatively higher than the lifespans of the leaves generated by leaf picking. This is because the leaves maintain their freshness for longer durations when they are still attached to the branches.
- The preservation and the maintenance of leaf quality become easier when the leaves are harvested via shoot cutting. It also becomes easier to feed the silkworms when the leaves are extracted in this manner.
- Another key advantage of shoot cutting is that the wastage of the mulberry leaves is minimized when this method is adopted.
- Finally, the maintenance of hygienic conditions also becomes easier when the mulberry leaves are harvested via Shoot cutting.

Apart from these points, it can also be noted that mulberry leaf harvesting via branch cutting is often associated with the highest leaf utilization by the silkworms.

2.5.3 Advantages of Shoot Cutting with Reference to Comparison to Leaf Picking

For the harvesting of mulberry leaves, shoots cutting is a more suitable method than leaf picking because of the following reasons:

- The leaf lifespans are relatively higher than the leaf lifespan produced by picking the leaves. This is because when already attached to the branches the leaves retain their freshness for longer durations.

- ✰ Furthermore, when the leaves are harvested through shoots cutting, the preservation and maintenance of leaf quality become simpler. Food for the silkworms often becomes simpler when the leaves are harvested in this way.
- ✰ Also, the shoots cutting method of mulberry leaf harvesting is less labour intensive, as it is easier to cut off the entire shoots than picking off each individual leaf. The labour costs associated with this approach are, therefore, much lower.

2.6 Fertilizer application in Moriculture

Dosage will depend on the fertility level of the soil, type of garden and density of plants. Fertilizers should be applied after six months of planting. The time schedule and quantity of fertilizers to be applied are:

Rain fed crop:	N100: P50: K50kg/ha/year.
June- August -	50:50:50 kg/ha (basal dose)
Sept -Nov. -	50 kg/ha (top dress)

There should be sufficient soil moisture at the time of application.

Irrigated crop:	N 300: P120:K120 kg/ha/year
After I harvest:	60: 60: 60
After II harvest:	60
After III harvest:	60: 60: 60
After IV harvest:	60
After V harvest:	60

Fertilizer application is done in five split doses and each dose is given ten days after harvesting of each crop. Fertilizers should not be applied immediately after harvest and also to the standing crop.

2.7 Preservation of Leaves

Mulberry leaves must be kept in a basket full of moisture. Basket should be lined with clean and wet cotton cloth. It should also be covered with wet and clean cloth. When preservation of mulberry leaves is done in bulk then preservation chambers are used. Moist and green mulberry leaves are suitable for feeding the silk moths. The leaves should be heaped loosely and periodically turned. By doing this aeration is provided to leaves for better preservation. Optimum temperature and humidity are required for preservation of mulberry leaves.

2.8 Uses of Mulberry

2.8.1 In Sericulture

The most important use of mulberry globally is in the production of silkworm, which feeds exclusively on its leaves. The country with the largest area of mulberry is China with approximately 626 000 ha, then India with nearly 280000 ha. Several other countries, such as Thailand and Brazil (35 000 ha), still have some mulberry production but on a much smaller scale. Silk projects have been and are being started in various developing countries, particularly in Africa and Latin America. Regardless of how successful or sustainable they have been, these projects have been responsible for the introduction and dissemination of mulberry varieties under different soil and climatic conditions.

2.8.2 As Fruit

All throughout Asia, but particularly in Central Asia and in the Near East, mulberry is highly appreciated for its delicious fruit, which is consumed fresh, or in the form of juice or conserves. A few mulberry orchards also exist in Latin America.

2.8.3 As Wood

Especially in the Indian subcontinent, mulberry wood is used for handicrafts, cabinet work and for sporting woods (*e.g.* grass-hockey sticks and tennis rackets). Its thin branches are woven into baskets.

2.8.4 In Landscaping

In Asia, southern Europe and in the southern United States, mulberry trees are utilized for landscaping (Tipton, 1994). Their resistance to pruning and their low water requirements make them very suitable plants for urban conditions, house gardens, street shade and city embellishment.

2.8.5 In Medicine

A variety of medicinal properties has been attributed to the different parts of the mulberry plant (Datta, 2000). Leaves are also dried and used in infusions in Asia (*e.g.* China and Thailand).

2.8.6 As Forage

Silk producers have traditionally fed mulberry surpluses from silkworm feeding to farm animals and to herbivorous carp in polyculture fish ponds (Gongfu, Zengqi and Houshu, 1997). In this sense, farm animals are well integrated into silk production. There have even been economic studies of the combined benefits from silk and dairy production in India (Mehla, Patel and Tripathi, 1987). A review of mulberry for animal feeding has been published by FAO (FAO, 1999). In countries south of the Himalayas, where mulberry trees occur in nature, their foliage has been part of the traditional mixed diet of domestic ruminants. In India there have been numerous research reports on the use of mulberry residues and leaves for various domestic animals, from cows to poultry. In general terms, Indian scientists have considered mulberry foliage as medium-quality forage. Some small farmers in East Africa, mainly in the United Republic of Tanzania and in Kenya, harvest foliage from mulberry trees and include it as part of the diet offered to ruminants in confinement. Although its nutritive value is recognized in these countries, there has been little planting of mulberry for this purpose. Since the late 1980s and mainly in the late 1990s, the cultivation of mulberry as forage has expanded considerably in Latin America, in Central America and in the Caribbean in particular, but it is still mainly associated with research institutions. This has been mainly the result of the initial studies carried out by the Tropical Agriculture Research and Higher Education Center (CATIE) in Costa Rica and in other Central

2.8.7 Mulberry Leaves Health Benefits

Overwhelming scientific evidence from research conducted at top medical institutions, prove that the mulberry leaf is one of the most powerful natural herbal supplements. The evidence is so powerful it can no longer be ignored as a potent solution to many health issues. The strongest research around mulberry leaf is related to its ability to reduce inflammation, as it is rich in many anti-inflammatory compounds. This may in turn help improve blood sugar and reduce cardiovascular risk factors. These incredible compounds found in mulberry leaf makes it an essential ingredient for helping lower the risk of diabetes and heart disease.

2.8.7.1 Mulberry Leaves and Blood Sugar

Diabetes is a debilitating illness that impacts over 10 per cent of the population. The powerful ingredient in mulberry leaf is called DNJ, which prevents the absorption of carbohydrates in r gut. When carbohydrates are not absorbed, this reduces blood sugar and insulin. The ability of mulberry leaf to reduce inflammation and normalize blood sugar may help reduce the risk of many chronic diseases, a critical benefit for public health.

2.8.7.2 Mulberry Leaves and Heart Health

Scientists have pinpointed several biologically active compounds in mulberry leaf that are effective in suppressing the progression of atherosclerosis, the buildup of plaque in our arteries. It does this by inhibiting the oxidation of LDL-cholesterol, a major factor in the development of atherosclerotic plaque. Mulberry leaf may also help reduce blood pressure and inflammation, two other contributors to heart disease. Some of the additional effects are:

- ☆ Contributes to weight management
- ☆ Supports normal function of the liver and kidney
- ☆ Suppresses mutagenesis of carcinogens
- ☆ Strengthens bone mass
- ☆ Improves skin
- ☆ Supports healthy digestion and absorption of nutrients
- ☆ Enriches the blood with nutrients
- ☆ Soothes nerves
- ☆ Helpful in treating constipation

Chapter 3

Mulberry Cultivation

3.1. Introduction

Sericulture is an agro-industry which means, combining both agriculture and industry, the final product is silk. It is a labor intensive activity is providing direct and indirect employment to a large number. India has the unique distinction of producing all the four varieties of silk *viz.*, mulberry, muga, tasar and eri. About 14 per cent of the global raw silk is produced in India, of which more than 50 per cent is mulberry silk, mostly produced in Karnataka. Besides meeting the ever-growing domestic demand, silk in its various forms is also exported and this sector is one of the important foreign exchange earners for the country.

Mulberry is derived from Latin word Morus and the cultivation of mulberry leaves of rearing of silk worms is called as Moriculture. Mulberry is grown as shrub in tropical countries and trees in temperate countries like Japan. It belongs to order urticales, the family Moraceae and genus Morus. There is an estimated about 68 species of the genus Morus with the majority of them occurring in Asia and in China there are over 1000 varieties under cultivation. The most common spices are M.alba, the white mulberry, M.nigra, the black mulberry and M.rubra, the red mulberry.

Mulberry cultivation is the agriculture part of sericulture which constitutes not only the rearing of silk worms but also silk reeling. Cultivation of mulberry plays a significant role in determine the production cost of cocoons and silk as it is estimated that 60-70 per cent of the cost of cocoons goes to mulberry. It is the chief food for Bombyx mori. Mulberry leaves protein is the source for silk worm to biosynthesize the silk, which is made up of two proteins *i.e.*, fibroin and sericin and its leaf and cell wall together contains with structural carbohydrates and which is highly digestible. So mulberry is the main food source for silk worms. One hectare of fertile land can produce about 15-40 tons of mulberry leaves over a 12-month period. This increase in the production of mulberry and reducing the cost of production. This increase in the production of mulberry and reducing the cost of production. Attention has been paid to intensive cultural operations including application of economic dosage of fertilizer and adoption of suitable irrigation schedules.

Distribution of mulberry cultivation on a global survey of sericulture industries reveals that there are at least 29 countries cultivation of mulberry plant is not only for silk rearing, but so many applications are used. Mulberry are widely distributed because of berries, some of the berries are very sweet tasting as honey, these syrupy sweet mulberries are used in ice cream, jams, jellies and beverages.

Beneficial health effects that may include the prevention of cancer and these mulberries having many medicinal values, due to its pigmentation it is used in preparation of stains, used in preparation of fabrics, the leaves of mulberry contain 6 times more calcium than green tea and 25 times more than spinach, mulberry leaves are also high in general content antioxidants and effective in regulating fat while boosting metabolism *etc.* Mulberry is the main food source for silk worm and the sericulture is mainly based on the mulberry for the production of silk. The superiority of the silk as a textile fiber has been recognized from the time immemorial; the luxurious look, sleek feel and luster of silk fabric are unquestionably inimitable. The industries have had an adventurous course of evolution, becoming established from time to time in other part of the world.

3.1.1 Leaf Chemical Composition

This differs according to variety and maturity. Mulberry leaves contain numerous chemical constituents. 1-Deoxynojirimycin (DNJ), phenolics and flavonoids are the prominent functional compounds. However, on the basis of the analysis carried out by CSRTI (Mysore), the chemical composition of the leaf is as follows:

Nutritional Value of V1 and G4 Mulberry Varieties (After the 6 hrs of harvest)

Plant Growth, Yield and Nutritional Parameters	*V1 Mulberry Varieties*	*G4 Mulberry Varieties*
Parentage	S30 x C776	*Morus multicaulis* x S13
Leaf Yield (mt/ha/yr)	>60	>65
Nitrogen (per cent)	3.98	3.94
Phosphorous (per cent)	0.333	0.375
Potassium (per cent)	1.79	1.64
Crude Protein (per cent)	24.88	24.63

Plant Growth, Yield and Nutritional Parameters	*V1 Mulberry Varieties*	*G4 Mulberry Varieties*
Total Sugars (per cent)	16.48	16.34
Moisture content (per cent)	75.8	75.3
Moisture Retention capacity (per cent)	80.2	78.6

3.2 Mulberry cultivation practices

3.2.1 Soil and Climatic Conditions

Mulberry can grow practically on any type of land except on very steep lands. Good growths, however, are obtained when it is raised on either flat land or gently sloping or undulating lands. On more slopy or steep lands, necessary attention to proper soil conservation methods as contour drains, contour planting or even bench terracing should be given.

Mulberry grows in a wide range of soils, but best growth is obtained in loamy to clayey loam soils. The mulberry plant can tolerate slightly acidic conditions in the soil. ln the case of too acidic soils with pH below 5, necessary corrective measures through application of Dolomite or Lime should be adopted. ln the case of alkaline soils, application of Gypsum should be resorted to, for correction of the soil alkalinity. Since mulberry is a deep-rooted plant, the soil should be sufficiently deep upto about two feet in depth. ln respect of elevation, mulberry thrives well upto about 4,000 feet, above which growth will be retarded because of the cooler temperature.

Mulberry cultivation can be done in a wide range of soils, but best growth is obtained in loamy to clayey loam soils. The mulberry plant can tolerate slightly acidic conditions in the soil. In the case too acidic soils with pH below 5, necessary corrective measures through application of dolomite or lime should be adopted.

In case of alkaline soils, application of gypsum should be resorted to for correction of the soil alkalinity. Since mulberry is a deep-rooted plant the soil should be sufficiently deep upto about 2 feet in depth. In respect of elevation mulberry thrives well upto about 4000 feet, above this level growth will be retarded because of the cooler temperature.

3.2.2. Land Preparation

Mulberry falls under the category of perennial crops and once it is properly raised during the first year, it can come to full yielding capacity during the second year and lasts for over 15 years in the field without any significant deterioration in the yield of leaf. Usually, flat lands are suitable for irrigated mulberry cultivation. If the slope is more than 15 per cent, suitable land development measures such as contour bunding, bench terracing should be adopted. Land for mulberry cultivation is ploughed deep with a heavy mould board plough up to a depth of 30-35 cm. Thereafter the land is repeatedly ploughed two or three times with a country plough to bring the soil to a fine tilth. The land should be properly levelled. A basal dose of well decomposed farmyard manure (FYM) or compost is applied at the rate of 10 tonnes/ha and thoroughly incorporated into the soil.

Weeds and stones should be removed during the preparatory stage. Basal dose of form yard manure at 20 tonnes per hectare is recommended, which as to thoroughly incorporate into the soil. Farm lands along the highways, in the vicinity of factories, area irrigated by raw sewage and untreated effluents plots affecting other gardens with intensive pesticide usage, water logged areas and tobacco grown land are not suitable locations for mulberry cultivation.

3.2.3. Suitable Mulberry Varieties

Based on the availability of water, resource and region the mulberry variety can be selected for cultivation

Kanva-2

It belongs to *M. indica.* Diploid. Widely cultivated in southern India after it was recommended for cultivation in 1969 by CSRTI (Mysore). Selection from natural population of Mysore local variety.

Inflorescence and sorosis: Female, profuse flowering, many soroses.

Production characteristics: Medium leaf maturity, yields 30-35 tonnes/ha/ year under irrigated conditions. Leaf moisture content 70 percent, protein content 21 percent and sugar content 11.5 percent. Resistant to leaf spot. Moderately resistant to leaf rust and powdery mildew. High rooting ability (80 percent) and wide adaptability.

S-36

It belongs to *M. indica*. Developed at CSRTI and recommended during 1984. Evolved from Berhampore local by chemical mutagenesis. Cultivated in southern India. Moderate rooting ability.

Production characteristics: yields 38-45 tonnes/ha/year under assured irrigated conditions of South India. Moisture content 76 percent, CP 22 percent and carbohydrate content 28 percent. Tolerant to leaf spot and powdery mildew. Moderately susceptible to leaf rust and to tukra infestation. Most suited to young age silkworm rearing. Sapling plantation recommended because of moderate rooting ability.

S-54

Belongs to *M. indica*. Developed at CSRTI and recommended during 1984. Selected from Berhampore local by chemical mutagenesis (EMS). Recommended for assured irrigated conditions of South India.

Production characteristics: yields about 45 tonnes/ha/year under assured irrigated conditions. Highly responsive to agronomical input. Moisture content of leaf 70.5 percent, CP 23.9 percent and sugar content 13.8 percent. Moderately resistant to powdery mildew and leaf rust and resistant to leaf spot. Loses moisture very quickly. Good rooting ability.

DD

Selected from natural population of Dehra Dun variety and recommended by Karnataka State Sericultural Research and Development Institute, Thalaghattapura. Recommended for southern India.

Morphology: erect, thin branches, coarse leaves, greenish grey bark. Lower branches spreading, leaves unlobed, big size, ovate shape.

Yields: 35-40 tonnes/ha/year under assured irrigation conditions.

MR-2

It belongs to *M. sinensis.* Diploid. Selection from open pollinated hybrid population. Developed at CSRTI and recommended for propagation in Tamil Nadu. Mainly cultivated in Tamil Nadu under both irrigated conditions in the plains and rainfed conditions in hilly regions.

Production characteristics: yields 30-35 tonnes/ha/year under irrigated conditions of Tamil Nadu. Moisture content 68 percent, protein content 23.2 percent, sugar content 13.2 percent. Resistant to powdery mildew disease. Suitable for hilly areas.

Victoria-1 (V-1)

It belongs to *M. indica*. Recently developed from a cross of S-30 and Berc 776 at CSRTI. Recommended during 1996 for assured irrigated conditions.

Flower: male, profuse flowering, occasionally few soroses.

Production characteristics: yields about 70 tonnes/ha/year under assured irrigated conditions. Very high sprouting. Moisture content 78.9 percent and 72.5 percent in young and matured leaves respectively, protein content 24.6 percent and total sugar content 16.98 percent.

Moderately resistant to leaf rust and tukra infestation and resistant to leaf spot. Quick sprouting ability and very high rooting ability (> 94 percent) high photosynthetic rate and higher water use efficiency are additional advantages. Moreover, leaves are suitable for both young and grown bivoltine silkworm rearing.

Sl.No.	*Variety*	*Leaf Yield (MT/ha/yr)*
	Irrigated condition	
1	G-4	60
2	G-2	
3	Vishala	55-60

Sl.No.	Variety	Leaf Yield (MT/ha/yr)
4	M-5	35
5	S-36	40-45
6	V1	60
7	DD	40-45
8	S-54	45
	Rainfed condition	
1.	S-13	18-35 (Suitable for Red soil)
2.	S-34	15 (Suitable for Black soil)
3	RC-1	25 (Resource constraint)
4.	K2 x Kosen (Sahana)	25 (Shade variety)
5.	AR-12	25 (Alkaline resistant)
6	RFS-175 and RFS-135	45
	Eastern and North eastern India	
1.	Ber S1	30-35 Irrigated condition; 12 Rain fed condition
2.	Ber S 799	35 Irrigated condition
3.	S 1635	40-45 Irrigated condition
4.	BC-2-59	10-15 Hilly area
5.	Tr-10	25 Hilly area
	North India	
1.	S146 (Check majra)	30-35
2.	China White	13-14 (bush), 8-9 (tree)
3.	Goshoerami	16-17 (two seasons) (tree); 25 (bush) rainfed
4.	KNG	16 (two seasons) (tree) 24 (bush) (rainfed)

3.2.4. Spacing

In cultivation, spacing depends upon the soil conditions, slope, variety and convenience for inter-cultural operations has detailed below in the case of rain fed mulberry gardens.

The aim should be to raise mulberry plant with a sturdier frame so that it is able to withstand prevailing drought conditions.

Commonly followed Spacing

Spacing (cm)	*Suitability*	*No. of Plants/ha*
90 x 90	Black soils, Heavy rain fall areas	12345
120 x 60	Hilly/Sloppy land	13888
(90+150)x 60	Paired row/Mechanized farming	13888

3.2.5. Planting Methods

Pit system is ideal in 90×90cm spacing. Pits of 3cm cube are dug and left for one month, which are later filled with FYM and soil at 1:2 ration. Trenchers of 35cm×35cm are also convenient to take up plantation in mulberry cultivation.

It is preferable to start the plantation during the rainy season. In highly areas, three cuttings of 20-22cm length are planted in each pit at a spacing of 15cm. cutting not sprouting in 4-5 weeks need to be replaced by saplings to ensure the required plant density.

3.2.6. Weeding and Inter-Cultivation

During the initial stage of mulberry plant established in the field, weed growth should be kept to the minimum, so that the growing young plants are not smothered by the weeds.

At least 2 weeding should be carried out during the first 6 months after planting of cuttings, once after 2 months of planting and again after an interval of 2-3 months. The weeding operations should be thorough and the soil should be dug deep to eliminate the weeds with roots. This deep digging is carried out as part of the weeding operation and results in necessary loosening of the soil and stimulation to the plants to grow vigorously.

Special care should be taken to reduce the weed growth as much as possible in the first year of planting. Thereafter, the shade effect of the fully grown in mulberry will tend to keep the weeds down. Similarly, periodical intercultivation helps in catching the rain water and its deep penetration for better retention of soil moisture.

3.2.7. Pruning and Leaf Harvest

In order to get good leaf yield for 15-20 years, it is necessary to give a good and study frame to the plant. After the planting, the saplings need to be cut at 10-15 cm above the ground level within a week. After the growing period of 6 months, the plants are to be cut a height of 25 cm above the ground level keeping 3-4 strong branches.

In leaf picking method, the first bottom pruning has to be taken up at a height of 20-25 cm from ground level and second bottom pruning is done by sharp sickle or pruning saw again at 30 cm height cm has to be followed once in a year during Jan to Feb and it in short harvesting method, each shoot harvest done at 25cm height which itself acts as pruning.

3.2.8. Manures and Fertilizers

Application of a basal dose of organic manure like compost or cattle manure is necessary for successful establishment of the garden. Thereafter, the young growing plants should be assisted to put forth vigorous and maximum growth through periodical fertilizer applications.

FYM has to be applied at 20t/ha/year in two doses following the first bottom pruning and third pruning. Fertilizers has to be applied has per the recommended schedule and secondary and micronutrients have to be applied wherever necessary.

Foliar sprays such as Boron (1 per cent) urea (0.5 per cent), zinc sulphate (0.1 per cent) will improve the leaf quality.

The recommended fertilizer dose (kg/ha) is indicated below:

Based on the soil test result, fertility level of the soil, method of leaf picking method the Fertilizers should be applied. The time schedule and quantity of fertilizers to be applied are:

- ☆ **Rain fed crop:** N100: P50: K50kg/ha/year.

  June- August - 50:50:50 kg/ha (basal dose)

  Sept -Nov. - 50 kg/ha (top dress)

  There should be sufficient soil moisture at the time of application.

- ☆ **Irrigated crop:** Application of chemical fertilizer in irrigated crop depends on hybrid and method of mulberry harvest fallowed to rear silkworm. The dose of chemical fertilizer and interval of fertilizer application for irrigated mulberry and different method of harvest given below

  N 300: P120:K120 kg/ha/year

  After I harvest: 60: 60: 60

  After II harvest: 60

  After III harvest: 60: 60: 60

  After IV harvest: 60

  After V harvest: 60

The recommended dose of chemical fertilizer for different method of mulberry harvest (5 shoot harvest/year or 6 leaf harvest/year) is given in the table.

Harvest	*FYM (Metric ton)*	*Fertilizer Dose (N:P:K kg/ha/year)*	
		Shoot Harvest	*Leaf Harvest*
After I harvest:	10 MT	70:28:28	60:60:60
After II harvest	-	70:28:28	40:00:00
After III harvest	-	70:28:28	40:00:00
After IV harvest	10MT	70:28:28	60:60:60

Harvest	*FYM (Metric ton)*	*Fertilizer Dose (N:P:K kg/ha/year)*	
		Shoot Harvest	*Leaf Harvest*
After V harvest	-	70:28:28	40:00:00
After VI harvest	-	-	40:00:00
Total	**20MT**	**350:140:140** (*Recommended dose for V1 mulberry)	**280:120:120**

3.2.9. Irrigation/Water Supply

Regular irrigation at an interval of 8-10 days is ideal. It may be noted that there is only one irrigation channels for every two rows of mulberry plants.

This helps in both saving and more effective use of the irrigation water. In case of water scarcity, drip irrigation can also be adopted for better production.

3.2.10. Harvesting of Mulberry Leaves

Leaves are harvested either by individual leaf harvest or shoot harvesting. The latter is more economical and used for the shoot method of silkworm rearing.

Picking of leaves should be carried out in time in mulberry cultivation, that is to say, when the leaves are the correct stage of maturity for harvest. Otherwise, part of the leaves become over mature coarse and suffer in quality from the point of view of the nutritive value for the silk worms.

Also, the part of the leaves may turn yellow, shed and be lost. Therefore, timely harvest, as the leaves reach the required stage of maturity, will lead to fuller harvest of the available leaves without wastage, and realization of maximum yield.

3.2.11. Yield in Mulberry

Cultivation normally the expected annual yield is 40-50 tone of leaves per acre. In shoot harvest method, harvesting can be done at an interval of 70 days (5 harvests).

In case of leave method, first harvest is taken 70 days after pruning and second and third harvest (coinciding with second bottom pruning) at an interval of 55days. The fourth harvest is taken 70 days after the third harvest and 5th and 6th at an interval of 55days.

3.3 Health Benefits of Mulberry

Mulberries contain dietary fibre, can help to improve digestive health by bulking up the stool, also reducing occurrences of constipation, bloating and cramping. Also helps to regulate cholesterol levels and can boost heart health when regularly added to the diet.

Mulberries contain carbohydrates that convert sugar into glucose. There by providing energy to the cells.

Mulberries increases iron intake and ensures sample supply of oxygen to the tissues.

Mulberries are rich in vitamin-K and vitamin-C, Increases tissue strength and boosts collagen synthesis. Vitamin-K helps in bone tissue development and is an essential component for blood clotting.

They also contain riboflavin(vitamin-B2) which protects tissues from free radicals and helps in transferring oxygen throughout the body

Mulberry are source of antioxidants, help lesser the damage caused by free radicals v Mulberries contain alkaloids that activate macrophages, which stimulate the immune system. v Mulberry has a remedy for swelling and redness.

Mulberry can nourish and promote production of body fluid.

Mulberries are rich in antioxidants and Phyto -nutrients that help in stopping the growth and spread of tumor cell they also contain resveratrol, which has anticancer properties, which fight against colon cancer, skin cancer, prostate cancer and thyroid.

Mulberries control blood pressure and cleans the blood, antioxidants help in improving the function of the blood vessels by keeping them supply and dilated.

3.3.1 Nutrition Facts

Fresh mulberries consist of 88 per cent water and only have 60 calories per cup (140 grams). By fresh weight, they provide 9.8 per cent carbs, 1.7 per cent fiber, 1.4 per cent protein, and 0.4 per cent fat.

Mulberries are often consumed dried, similar to raisins. In this form, they contain 70 per cent carbs, 14 per cent fiber, 12 per cent protein, and 3 per cent fat — making them fairly high in protein compared to most berries.

Main nutrients in a 3.5-ounce (100-gram) serving of fresh mulberries are calories- 43, water 88 per cent, protein 1.4 grams, carbs- 9.8 grams, sugar- 8.1 grams, fiber- 1.7 grams and fat- 0.4 grams.

Mulberries contain high amounts of both iron and vitamin C, as well as decent amounts of potassium and vitamins E and K.

Chapter 4

Pests and Diseases of Mulberry

Different kinds of disease of mulberry which are caused by fungi, bacteria, viruses and also mineral deficiencies are physiological disorders. Of these, 10 or 12 diseases cause much damage to mulberry plants. They are widespread in all the mulberry growing countries of the world. The diseases affect different parts of the plant. The mulberry being a perennial crop the pathogens seem to readily perpetuate and quickly spread to cover extensive areas. Certain alternate and collateral host (mainly weeds) places some role in perpetuation of some of the pathogens.

4.1 Fungal Diseases

4.1.1 Alternaria Leaf Blight

Causal Organism: *Alternaria alternata*

(Systematic position: Kingdom-Fungi, Division-Deuteromycota, Order-Hyphomycetales,

Family- Dematiaceae, Genus- Alternaria, Species- alternate)

- ✰ The disease is common in southern part of India. The disease causes severe defoliation and mortality during summer and rainy season.
- ✰ Alternaria belongs to the division Deutero mycota. The spores of *Alternaria alternata* are multi-celled and pigmented and they are produced in chains or branching chains.

- ☆ The spores have a distinctive appearance that makes them easy to recognize.
- ☆ They are broadest near the base and taper gradually to an elongate beak.

Symptoms

- ☆ Burning of leaves at the leaf tip and become brownish.
- ☆ It manifests by the appearance of burning at leaf tip initially and spread towards the middle which later becomes brownish, black. Infected leaves dry up within 10 days and ultimately mulberry seedlings die after few days.
- ☆ Appearance of isolated irregular dark brown patches can seen on leafs
- ☆ Crop loss due to this disease is 10-12 per cent

Description of the Pathogen

- ☆ The genus Alternaria is characteristic by the large conidia. They are produced on chains, light brown to brown, ovoid or obclavated and divided by transverse and vertical walls, with or without beak at the tip.
- ☆ The hyphae and conidiophore are light brown and septate. Generally, the maximum width of spore is from 5-13 to 6-24 μm (the mean is 9 to 17 μm), and the length of spore is from 16-37 to 35-110 μm (the mean is 37 to 69 μm).
- ☆ The number of transverse septa is from 0 to 23. The beak length is from 4 to 90 nm. The conidiophore length is from 20 to 300 μm. Usually, the conidia formed in natural habitats are larger, have longer beaked tip, and are more uniform than those formed in common agar media.
- ☆ Under low temperature and dry conditions, larger conidia with shorter beak are produced. On the culture it is grey to Under brown, downy or cotton colony.

Pre Disposing Factors

- ☆ Alternaria blight is worse in late summer because longer periods of dew and higher relative humidity above 95 per cent are recorded in August and September than in June and July. At the same time, aging leaves are more susceptible. High relative humidity and dew favor the disease.
- ☆ The disease starts on leaves of the exposed shoots on the top of trees and eventually moves towards the lower parts of the tree.

Disease Cycle

- ☆ Alternaria survives winter as mycelium or spores in infected plant debris left on or in the ground, under favorable conditions in the spring, the spores quickly mature into fungal leaf spots that release more spores carried to other plants by wind and splashing rain.

- ☆ This cycle repeats throughout spring and summer for as long as favorable conditions persist.

Cultural control: Removing plant debris from garden plots each fall and then tilling the plot. Prune off the over matured shoots.

Fallowing paired row system of planting which hinders the incidence level

Chemical control: Apply 0.2 per cent Indofil M45. A second spray may be done if the disease is more severe. The leaf can be used for silkworm rearing 14 days after the fungicide spray.

4.1.2 Leaf Spot

Causal Organism: *Cercospora moricola*

(Systematic position: Kingdom–Fungi, Division-Ascomycota, Class-Dothideomycetes, Order- Capnodiales, Family-Mycosphaerellaceae, Genus-Cercospora, Species-moricola)

- ☆ The leaf spot caused by Cercopsora to mulberry was first reported by Patel *et al.* (1949) from Kamataka. The disease is reported to cause 10-12 per cent leaf yield loss even up to 20 per cent -35 per cent in severe conditions.
- ☆ Maximum severity was found during rainy season fallowed by winter season in Kamataka. The disease causes reduction in moisture, proteins and sugars in infected leaves.
- ☆ Rearing with infected leaves affects health of silkworms and in turn quality and quantity of cocoons produced.
- ☆ 35-40 Days after pruning the disease

Symptoms

- ☆ The causal fungus produces minor circular light brown spots on the leaves which gradually increase in size and turn darkbrown. The tiny spots coalesce to form bigger ones.
- ☆ The dead tissues fall off resulting shot holes lining yellow circle around the shot hole. The highly affected leaves fall off prematurely

Description of the Pathogen

- ☆ The conidia are born on conidiophores. Conidiophores 2-15 in a fascicle, pale olivaceous brown, straight, rarely septate and geniculate, unbranched, scars conspicuously thickened,10-50 x 4-5.5 μm.
- ☆ Conidia are hyaline, solitary, acicular, multi-septate base-truncate, tip-acute hilum thickened and darkened, 40-150 x 2-3.5μm.

Pre Disposing Factor

- ☆ The disease is generally found during rainy season followed by winter. High disease development was recorded in the month of August and September when outdoor temperature and relative humidity are 25-30 °C and above 80 per cent respectively.
- ☆ The disease starts progressing 35-40 days after pruning (DAP)/leaf harvesting and becomes severe 70th day after pruning.

Disease Cycle

- ☆ The disease is very common during June –December, conidia germinate on both surfaces of the leaves. However, germ tubes (single or branched) enter through stomata only whether open or closed, indicating the absence of direct penetration.
- ☆ In mulberry, the stomata are distributed only on the lower surface of the leaves (hypostomatous). Therefore, successful infection of *C. moricola* is possible only when the conidia are deposited on the lower surfaces. Interestingly, conidia are carried to the undersurface by rain splashes ensuring natural portal for entry and assured moisture for germination and penetration.
- ☆ The conidia are capable of producing new hyphae from any cell. It takes about 10-12 days after inoculation of conidia to produce a spot and another 3-4 days for the production of conidia.

Control Measures

Cultural Control

- ☆ Follow wider spacing of plantation (90 cm x 90 cm) or paired row planting system [(90 +150) × 60 cm]. Avoid sprinkler irrigation.
- ☆ Use of resistance variety for plantation

Resistant Varieties

Resistance Reaction	*Name of the Variety*
Completely resistant	Kalliakutahi, Bilidevalaya (Govindaiah *et al.,* 1989)
Resistant	Goshoerami, Ichinose,Kanva-2, S54, C799, Shrim-2, Shrim-8,MR-2, Assambola, AB x Phill, mouli, Mizusowa, English black, K2xKosen, China peking, Cattaneo, Sujapur x Kukuso-13, China white, Jodhpur, AB x Kokuso-13, Calabresa, Miuraso, Acc-153, Acc-152,Acc-151, Acc-150, Acc-128, Acc-135, Acc-106, Acc-109, Acc-112, Acc-114, Acc-115, Acc-116, Acc-117, Acc-119, Acc-121, Acc-123, Acc-124,OPH-1,OPH-3, MS-2, MS-5, Acc210, Paraguay, RFS-135. RFS-175, S-146, S-523, MS-7, MS-8, Acc125, Acc-128, K2 x Kokuso, S1531, LF-1, Almora local, S31, S1096.

Chemical Control: Various fungicides are identified for control of the disease.

- Spraying 0.2 per cent Bavistin (Carbendazim 50 per cent WP) solution on the leaves. This can be made by dissolving 800g Bavistin in 200-liter water for spraying in 1-acre mulberry garden. The leaves can be used for silkworm rearing 7 days after the spray.
- Spray 0.2 per cent Karathane. About 400-500 ml Karathane is to be mixed with 200-250 liters of water. A second spray can be made if the disease is more severe. The leaves can be used for silkworm rearing 10 days after final spray.

4.1.3 Fusarium Leaf Blight

- The disease is a minor disease first reported from Mysore. Though the disease occurs all over India, the incident is found more in south India.
- The disease is also known as twig blight. However, the twig blight was reported as a serious disease of mulberry in the Kashmir valley.

Causal Agent: *Fusarium pallidoroseum* (Cooke) Sacc

(Systematic Position: Division: Eumycota, Sub Division: Deuteromycotina, Class: Hyphomycetes, Order: Moniliales, Class: Deuteromycetes, Family: Moniliaceae, Genus: Fusarium, Species: pallidoroseum)

Symptoms

- Appearance of dark brown spot-on infected twigs, shrinkage of cortex, bark bursting, yellowing and pre-mature leaf fall are the main characteristic symptoms of the disease.
- Marginal browning of leaves and forming of irregular black lesion at the at the initial stage of infection. In the later stage, the lesions coalesce and spread longitudinally resulting in splitting and drying of branches.
- The branches become feeble and fragile. Bursting of bark, yellowing and pre-mature leaf fall are observed in the later stages.

Description of Pathogen

- Fusarium pallidoroseum grow on Potato Dextrose agar medium at 25-28°C and produce rose pink coloured woolly to cottony, flat, spreading colonies when mature. Hyaline septate hyphae with slightly bulged compartments, conidiophores, phialides, macro conidia ad micro conidia are found. The fungus also produces chlamydospores.
- Phialides are cylindrical with a small collaret, solitary or produced as complex branching system. monophialides and polyphialides are observed. Macroconidia (3-8 x 11-70 μm) are produced from phialides on unbranched or branched conidiophores. They are two or more celled, thick walled, smooth, cylindrical or sickle shaped.

- Macroconidia are of two types. Those born in the aerial mycelium are mostly straight, 3-5 septate, and measure 7.5-35 x 2.5-4 µm. Macro conidia born in sporodochia are curved, possess a foot cell, 3-5 septate measure 20-46 x 3-5.5 µm. Micro conidia (2-4 x 4.8 µm), on the other hand, are formed on long or short simple conidiophores. They are one celled (occasionally 2 or 3celled), smooth, hyaline, ovoid to cylindrical, and arranged in balls (occasionally occur in chains).
- Chlamydospores are sparse, present in pairs, clumps or chains. They are thick walled, hyaline intercalary or terminal, (5-8 x 8-10 µm).

Disease Cycle

- The disease prevails throughout the year with peak in summer and rainy season and cause 5-8 per cent yield loss. Pathogen was present throughout the year on un-pruned blighted twigs with maximum conidial density from June to August and that of chlamydospores from December to March.
- There was positive and significant correlation of air temperature with conidial density and disease development. Dispersal of spores occurs by water splash, blow off and wash off mechanisms and also through insects.
- The pathogen perpetuated on diseased twigs in the form of mycelium, conidia and chlamydospores. The proportion of conidia decreased with increase in depth of placement of blighted twigs in soil.

Pre-disposing Factors

- The disease is air borne and spread through conidia. High humidity is the key factor for spread of the disease. The pathogen perpetuated on diseased twigs in the form of mycelium, conidia and chlamydospores.
- There is positive and significant correlation of air temperature with conidial density and disease development.

Control

- The application of Bavistin @ 0.2 per cent or Foltaf @ 0.2 per cent concentration is recommended for control of the disease.
- A safe period of 7 days is to be kept in case of bavistin and 10 days in case of Foltaf after which the leaves can be used for silkworm rearing.

4.1.4 Grey Leaf Spot

- Pseudocercospora mori the grey leaf spot pathogen of mulberry, has been reported from Japan, Congo, Taiwan, China, the USA and India.
- The fungus can cause significant damage to leaves of Morus spp. This is one of the major diseases during post monsoon season in the Gangetic plains of West Bengal and North-Eastern region.
- The disease appears during September and continues up to November.

Causal Organism: *Pseudocercospora mori*

(Systematic position: Kingdom- Fungi, Phylum: Ascomycota, Class: Dothideomycetes, Order: Mycosphaerellales, Family: Mycosphaerellaceae, Genus: Pseudocercospora Species: mori)

Symptoms

- The disease is characterized by small to medium size velvety gray to black spots appearing on the lower surface of the older leaves.
- The spots are indistinct on upper surface, more conspicuous on the lower surface where they are brown, angular or spreading (particularly along leaf margins) and restricted by the larger veins, up to 7mm wide. Later, the spots cover whole lower surface.
- Severely infected leaves turn yellowish and defoliate prematurely.

Description of the Pathogen

- Sporulation is hypogenous, ranging from olivaceous and velutinous to darker brown and punctate. Immersed hyphae pale olivaceous, smooth, septate, branching, 1–3 µm diam.
- Superficial hyphae sparse, emerging from stomata, pale olivaceous, smooth, septate, 1–2 µm diam., bearing lateral conidiophores similar to those associated with stomatal egress.
- Stromata lacking, reduced to a few light brown cells in the substomatal cavity, or compact and light brown, filling the sub-stomatal cavity, 18–25µm diam.
- Conidiophores are solitary or 2–22 in a divergent fascicle, olivaceous brown throughout or slightly paler at the apex, irregular in width, branched, straight, curved or sinuous, markedly geniculate towards a conic apex, conspicuously 1–4(–7) septate, 30–67 µm long×3–4.5 µm wide.
- Conidiogenous lociare not darkened, not thickened, non-refractive, 0.5–1 µm diam. Conidia are pale olivaceous, straight or curved, smooth, narrowly obclavate, straight to mildly curved, occasionally constricted, usually narrowing gradually to a sub-obtuse or obtuse apex, and a little more abruptly to an obconic, truncate base, 1–5(–7) septate, 18–61×3–4 µm. Hila are not darkened, not thickened, non-refractive, 1–2 µm diam. The pathogen grows in host extract agar medium.

Pre-dsposing Factors

- Temperature 25-30 and humidity 85-90 favoure the disease development.
- The disease infection was found higher during the growth period 48-55 days of the shoot.

Disease Cycle

- Since the low temperature favours the disease, the disease starts generally during the month of June with the onset of monsoon.
- The disease gradually develops with the lowering the temperature. The conidia disperse through the rain and during the month of September-October the disease will be very high and thereafter the severity will be low.
- The disease is generally not seen after the month of February due to the high temperature.

Resistant Varieties

- **Highly resistance varieties:** Thailand lobed,M. multicaulis, Italian, M. australis,
- **Resistant varieties***:* MS-8, Fernodias, MS-7, Thailand (lobed), PGDTR-9, Tomeiso, KNG, CSRS-II, Morus cathyana, Berhampore-B, Roznitul, Kurimoto, Kanmasari, Calabresa, Jodhpur, Morus australis, Thailand (unlobed), Berhampore-6, Assambola.

Control: The disease is effectively controlled by foliar spray of 0.1 per cent Carbendazim (Bavistin) - a broad spectrum systemic fungicide.

4.1.5 Black Leaf Rust

- Leaf rust, caused by *Cerotolium fici* also called as black rust is a very important foliar disease of mulberry in India.
- The disease generally appears during late winter and prolong up to early summer (September to February).
- Black rust generally appears on mature leaves and cause 5-10 per cent loss in leaf yield. Besides, the disease significantly reduces moisture, crude protein, sugars and total sugars in infected leaves.
- This disease progressing generally 35-40 DAP becoming sevear on 65 DAP. The old mature leaves become more prone to this disease
- The annual crop loss due to black leaf rust is 10-13 per cent

Causal Organism***: Cerotelium fici/Peridiospora mori***

(Systematic Position: Kingdom: Fungi, Division: Basidiomycota, Class: Teliomycetes, Order: Uredinales, Family: Phakopsoraceae, Genus: Cerotelium, Species: fici)

Symptoms

- The infected leaves show pin head like eruptive brown pustule on the lower side of the leaves and dark tiny angular spots on the concomitant upper surface of the leaves.

☆ As the disease become severe, leaves become yellow, wither off and fall prematurely.

Description of the Pathogen

☆ Hemi-form. Pycnia and aecia unknown. Uredia hypophyllous, scattered ± irregularly or in groups, sometimes on conspicuous angular spots between the veins, often dense and often profuse over part or all of the leaf, pale to ferrugineous, circular, 0,15-0,4 mm diam., like minute blisters open in the centre, often with inconspicuous cylindric-clavate, thin-walled, hyaline paraphyses.

☆ Urediospores globose, ellipsoid or obovoid, sometimes rather irregular and angular, 19-30 x 15-23 μm; wall hyaline to cinnamon, occasionally even darker, 1-1,5 (-2) μm thick, finely to strongly echinulate with spines 1-3 μm apart and 0,5-1 μm high; pores usually very obscure, but definitely 3 (-4?) and equatorial in some of the deeper coloured spores.

Predisposing Factors

☆ The disease is highly influenced by environmental factors, host age and genotype. The low atmospheric temperature (22-26°C) and high humidity (70 per cent) following monsoon (September-February) is highly favourable for the disease development.

☆ The leaves on the branches more than 45 days older are more prone to the disease and the younger leaves are generally free from the disease.

☆ The pathogen perennates through the dried leaves and plant parts found in the mulberry gardens.

☆ The pathogen Cerotelium fici is an obligate microcyclic rust fungus. It exists primarily as mycelium, uredium and uredinospores. The uredinospores are oval to round produced singly on uredophores in uredia.

☆ In favourable conditions for the growth of pathogen is 22-24°C temperature and high humidity favour the disease. Uredospores germinate and produce hyphae which enter the leaf through stomata.

☆ The hypha grow inter cellular in the host tissues sending haustoria in to the host cells to draw nutrients. Uredospores disperse through water droplets and wind currents and spread the disease.

Control

☆ **Cultural control:** Adopt wider spacing, do not allow the plants to over mature and prune off the whole plant immediately after rearing. Keep the mulberry garden clean and remove the dried leaves from the garden and burn.

☆ **Resistant varieties**

☆ **Resistant: C1726, S700, China Peking, Cattaneo**

- ☆ **Moderately resistant:** AB x Phill,K2 x Kosen, ACC115, Almora local, MS7, S13, MS2, MS6, ACC217, S34, ACC201, S25.
- ☆ **Chemical control:** Spray Kavach (Chlorothalonil 75 WP) 0.2 per cent *i.e.* dissolve 3 gram Kavach in 1 lit. water and thoroughly drench the leaves while spraying the fungicide.
- ☆ **Safe period:** The leaves can be used for silkworm rearing 14 days after the fungicide spray.

4.1.6 Powdery Mildew

- ☆ Powdery mildew caused by *Phyllactinia corylea* is a most common, serious and wide spread disease of mulberry. The disease is reported from different countries.
- ☆ In India, it was first reported by Ramakrishnan and Sudan (1954) from Madras. The disease is prevalent during September to March with maximum incidence in January and February. The disease reduces protein.
- ☆ The infected leaves are characterised by less moisture and disrupt the metabolic activities of the host leading to reduction in leaf yield and nutritive value.
- ☆ The incidence begins from 45th days after pruning and peak level reaches on 70th days after pruning

Causal Organism: *Phyllactinia corylea/Phyllactinia moricola*

(Systematic position : Kingdom-Fungi, Division-Ascomycota, Order-Erysiphales, Family-Erysiphaceae, Genus-Phyllactinia, Species-corylea)

Symptoms

- ☆ White circular powdery patches appear on the lower surface of the leaves. The concomitant upper surface develops chlorotic lesions.
- ☆ In the later stage, the white powdery patches turn to brownish-gray black; leaves turn yellowish, leathery and defoliate prematurely.

Description of the Pathogen

- ☆ The vegetative hyphae of Phyllactinia corylea produce a large number of long hyaline septate conidiophores on which conidia measuring 20 x 70 μ are borne. The pathogen differs from other powdery mildew fungi in having germinating cells terminally while the others have sub-terminally.
- ☆ The mycelium may be abundant and persistent, or scant and short-lived (evanescent). The cleistothecia are large (216–245 μm), with soft wall tissue, and obscure cellular structure and cracks and wrinkles (reticulations) showing appendages with bulbous base.

- ☆ The cleistothecia typically develop 8–12 easily detachable hyaline appendages that vary in length from 191–290 µm long. The asci are 4 to 5 to 20 or more, ovate, supported by small stalk-like structures (pedicellate), with dimensions of 72–83 by 32–40 µm. They are typically 2 per ascus, sometimes 3 or 4, and they are 31–36 by 21–25 µm.
- ☆ The filaments can gelatinize by absorbing water and are thought to function in helping the ascomata adhere to the surface on which they grow, like the underside of leaves. The feet are cylindrical, irregular in width, 32–72 by 7.5–25 µm, and divided into 2–10 branchlets in the upper part. Each branchlet is short, bulbous, with filaments being 20–42 µm, somewhat shorter than the foots, which are 2–4 µm wide.

Pre-disposing Factors

- ☆ Temperature 24° to 28°C is optimum for conidial growth.
- ☆ Conidia begin germinating within two hours at 28°C and 100 per cent humidity and the disease appears two weeks after inoculation.

Disease Cycle

- ☆ Dissemination of conidia takes place through air current. The fungus perennates unfavourable conditions through sexual fruiting bodies known as cleistothecium. These are flat, sphere-shaped, dark-coloured structures having several asci.
- ☆ Development of cleistothecia takes place at low temperature (14 to 17°C average day temperature) and moderate relative humidity fluctuating from 60 to 70 per cent during the months of December and January on living intact leaves.
- ☆ Initially they appear as tiny orange-coloured oval bodies which turn brown to black in due course. These act as a primary source of inoculum under favourable conditions.

Cultural Control

- ☆ Follow wider spacing of plantation (90 cm x 90 cm) or paired row planting system [(90 +150) × 60 cm].
- ☆ Avoid stagnation of water and shade in the mulberry plantation.

Resistant Varieties

- ☆ Immune: Kaliakutai
- ☆ Resistant: Mandalaya, Cattaneo, China white, Jodhpur, Calabresa, Mizusowa, Acc-123, Acc 125, Acc-151, Acc-153, OPH3, S523, Punjab local, Shrim-2, Himachal local, S796, S1531, S1096, S31, Almora local

Biological Control

- ✰ Hyphomycetes fungi, Cladosporium oxysporum and Cladosporium spp., have been reported to be hyperparasitic on P. corylea. Besides, Reddy *et al.* (1989) reported a coccinellid insect, *i.e.* Illeis cincta, eating the mycelium and spore mass of powdery mildew.
- ✰ Another species of this insect (I. indica) has been found to feed very commonly upon spores, mycelia and young cleistothecia in West Bengal, India.

Chemical Control

Chemical control: Spray 0.1 per cent Carbendazim 50 per cent WP (Bavistin) [2 g/lit. water] or 0.2 per cent Sulphur 80WP (Sulfex) [2.5 g/lit.].

4.1.7 Phloeospora Leaf Spot

Introduction

- ✰ This disease also known as brown leaf spot occurs in almost all sericulture areas of the world. In India, the disease was first reported by Munshi *et at.* (1986) from the Kashmir valley.
- ✰ The disease occurs during rainy to autumn season (June – October). The fungus causes heavy damage to the leaves, rendering them unsuitable for silkworm rearing. None of the mulberry cultivars grown in the valley was reported free from the disease.

Symptoms

- ✰ Infection starts with presence of small dark brown, then light brown later on yellowish or whitish; circular spots on both surface of the leaves.
- ✰ The spots turn grey in the centre surrounded by a broad yellow diffusion zone. The necrotic portion of the spots decay in the wet season and fall off and hence leaves will become perforated.
- ✰ During the dry season the necrotic portion split open. The spots at maturity become yellowish grey.

Causal Organism: *Phloeospora maculans*

(Systematic Position: Division: Ascomycota, Order: Dothideomycetes,Class: Pleosporales,Family: Pleosporaceae, Genus: Phloeospora, Species: maculans)

Pre-disposing factors: The disease is air borne and spreads through conidia. High temperature and humidity especially high humidity is the key factor for severity of the disease. Therefore, the wet condition in the mulberry field results in high incidence and severity.

Disease Cycle

- ✰ The pathogen overwinters in the form of pseudothecia or conidia on the leaves the plant remaining attached to mulberry trees or on decomposed organic matters. In the soil.

- ☆ The conidia disseminate in the wind and rain. The new infection will happen within 10 days and hence the re-infection occurs many times in a season.

Control: Spray Bavistin (0.05 and 0.1 per cent) or Captan (0.4 per cent) for effective control of the fungus. A safe period of 7 days in case of Bavistin and 14 days in case of Captan is required before using the fungicide sprayed leaves for silkworm rearing.

4.1.8 Red Rust

- ☆ Red rust of mulberry rust, caused by Aecidium mori Barclay, occurs on buds, shoots, leaves, leaf pedicels, flowers and fruits of mulberry plants. It causes a deformation of young leaves and heavy defoliation of affected mulberry plants. Red rust is widely distributed in China, Japan and Korea.
- ☆ In India, the disease is found mainly in the high altitude area such as Simla (Himachal Pradesh), Darjeeling (West Bengal) and Shillong (Meghalaya).
- ☆ Although the affected leaves are not toxic to silkworms, the cocoons are inferior and thus cause loss in cocoon production.
- ☆ The occurrence of this disease is noticed during rainy and winter season

Symptoms

- ☆ The affected part of young bud become swollen the young bud curls up in the abnormal shape and many densely and slightly protruded yellow spots are found on the malformed bud, which are the aecidia of the fungus.
- ☆ On both surface of the affected leaf, there are numerous round shiny spots which later protrude gradually turning yellow.
- ☆ Finally, the epidermis on the yellow spot break and yellowish powdery substances the aecidiospores scatter everywhere on the leaf.

Causal Organism: *Aecidium mori*

(Systematic Position: Kingdom- Fungi, Phylum- Basidiomycota, Class- Pucciniomycetes, Order -Pucciniales, Family- Pucciniaceae, Genus: Aecidium, Species: mori)

Pathogen Description

- ☆ The mycelia masses are formed under the infected portion of the epidermis. This mass gradually develops and protrudes like yellow pustules. These structures gradually mature turning from light yellow to dark yellow in colour. This erupts through the epidermis of the host plant.

- ☆ The acia occur and their bell-shaped opening mostly on the upper surface of the leaves. Generally, the aecia are about 150μ in diameter. It is surrounded by a layer of protective membrane made of oval, polyhydral cells, on the surface of which are found small spines.
- ☆ On the base of the aecium are rows of colour less cylindrical conidiophores (30x5 μ) on top of which are born chains of aeciospores. The aeciospores are colourless and polyhedral at the beginning and later become circular in shape. When mature, the aeciospores (13-20x10-17 μ) are circular or oval, orange in colour.
- ☆ They disperse through opening of the aecium (200 μ), germinate quickly under adequate temperature and humidity and grow within three days. However, vitality of aeciospores is affected by the freshness of the mulberry leaves and environmental conditions.

Predisposing Factors

- ☆ The occurrence of the disease is directly related to climatic conditions. Basically this disease is airborne it spreads by aeciospores through rain splash. The temperature range for the outbreak of disease is found to be 10-27°C.
- ☆ The most favorable temperature for the growth is 20-25°C. Within the range of this optimum temperature, humidity plays an important role during the infection. If the relative humidity is more than 80 per cent, the rate of infection will be higher and if the relative humidity is 77-78 per cent the rate of infection will be low.
- ☆ Similarly, when the temperature ranges between 18-30°C, the rate of infection would increase as the temperature rises, but the formation of aeciospores would be inhibited in a temperature higher than 30°C. At the temperature above 30-39°C the disease will not occur.

Disease Cycle

- ☆ Mycelia of the fungus overwinter on the branches of mulberry. During the following spring, aecia are formed as the bud sprouts and aeciospores are disseminated by rain and wind.
- ☆ After the germination of aeciospores, the tips of the germ tubes adhere to the hosts epidermis penetrate in to the cuticle and the epidermal cells develop in to mycelia.
- ☆ These spread in to the hosts tissue and produces haustoria which will absorb nutrients from the host cells.

Resistant Varieties: The mulberry varieties such as Alfonso, Kanmochi, Lun 104 are reported resistant to the disease.

Control

- ☆ 0.25 per cent wettable powder of 5 per cent zinc or 075 solution of 5 per cent wettable powder of nitrite are reported effective for controlling the disease.
- ☆ The leaves can be used for feeding after a waiting period of 7 days. If the disease is more severe, 1 per cent bleaching powder solution should be sprayed all over the plantation.
- ☆ Dusting of 0.2 per cent sulpher powder on mulberry leaves is suggested for effective control

4.1.9 Tar Spot

- ☆ The disease was first reported in mulberry by Heming *et al.* (1988) from Suzhou Sericulture Institute, China and Govindaiah *et al.* (1989) from India.
- ☆ The disease appears in the form of large irregular, circular or linear black spots on the leaves. The disease causes huge loss in severe condition.
- ☆ It is most prevalent in the mulberry growing areas of West Bengal and north Eastern states in India especially during the rainy season with warm temperature and high humidity.

Causal Agent: *Myrothecium roridum*

(Systematic Position: Class: Ascomycota, Sub Class: Sodariomycetes, Order: Hypocreales, Family: Not assigned, Genus: Myrothecium, Species: roridum)

Symptoms

- ☆ The spots are circular at first enlarging progressively and become irregular in the later stages especially under extensive moist condition. The vein tissues are included in the necrosis.
- ☆ The circular spots vary in size up to 2 cm dia., occasionally centre of the lesion cracks irregularly and rupture.
- ☆ Annual crop loss due to red rust disease is 10 per cent

Pathogen Description

- ☆ Sporodochia on leaf are sessile and polymorphic in surface view, Spore mat wet when young, shinny, black, convex surrounded by a fringe of entangled white hyphae.
- ☆ Marginal hyphae contorted, hyaline usually tapering towards a blunt apex, branched septate cells 20-30x 1-2.2μ. No external mycelium found on plant material. Conidiophore arises directly from the mycelium. Conidiophore hyaline, cylindrical, branched below and then in to 2-3 branches bearing phialides, cosely compacted to form a sub-hymeneal layer septate, cells 6.3-23 (14.7) x 1.4-3.6 (1.9) μ.

- ☆ Phialids in whorls of 2-5 at the apex of conidiophores branches, mostly cylindrical, hyaline, some dilute olivaceous, 9-27 (14.2) x 0.9-1.8 91.4) μ, closely compacted in to parallel rows forming a dens hymeneal layer. Spores rod shaped ends rounded, one end rarely slightly truncate, smooth walled, hyaline to olivaceous, black in mass, guttulate, 4.5-10.8 (7.2) x 1.3-2.7 (1.8) μ.
- ☆ Sporulation in culture often in concentric zones and consists of small group of conidiopores, formimg rudimentary sporodochia. The fungus grows well in potato dextrose agar medium with excellent sporulation.

Pre-disposing factors: Warm moist weather is conducive for the disease. The disease spread through the rain water.

Disease cycle: The pathogen survives in the soil. With the advent of rain, the spores will be carried along with the rain droplets causing infection.

Resistant varieties: Kajli, Jatinuni, Morus cathayana, Almora local, Bogura -1, Meergund-6, Fernodias, Punjab local, M. tiliafolia, Sulthanpur, Golaghat, Bush malda-A, Sujanpur.

Control

- ☆ **Biological control:** Few bacterial species such as Bacillus lentimorbus, B. cereus and B. pumilus are reported to control the disease.
- ☆ **Chemical control:** Apply 0.2 per cent Bavistin (Carbendazim (50 WP). Completely drench the plants with the fungicidal solution. The leaves can be used for silkworm rearing 7 days after spray.

4.1.10 Sclerotium Wilt/Collar Rot

- ☆ Wilt of sprouted mulberry cuttings caused by Sclerotium rolfisii was reported first by Siddaramaiah and Patil (1984) from Dharwar, Kamataka (India).
- ☆ The disease is very common in the rainy season and cause wilting sprouted cuttings in mulberry nursery. The disease is common in the established mulberry plantations also.

Causal Organism: *Sclerotium rolfsii* Sacc.

(**Systematic Position** :Kingdom: Fungi, Division: Basidiomycota, Class: Basidiomycetes, Order: Agaricales, Family: Typhulaceae, Genus: Sclerotium, Species: rolfsii)

Symptoms

- ☆ The disease is characterized by presence of dark brown to black water-soaked spots on the collar region of the stem which rapidly encircle the stem and eventually more upward and downward regions.

- ✩ White mycelial mat with a large number of chocolate brown sclerotial bodies will develop over the infected tissues.
- ✩ Sclerotia are ellipsoidal in shape and vary from 0.5 to 1.5 mm in size. Once the stem base is affected, the entire shoot drops down.

Causal Organism: *Sclerotium rolfsii* Sacc.

Description of the Pathogen

- ✩ The mycelia are white in early stages of development become tan as they mature and finally form strands of brown pigmented hyphae.
- ✩ They measure 6-9 μm wide have clamp connections and numerous dikaryotic nuclei. The sclerotia resemble mustard seeds, are round to irregular in shape, generally tan outside and have a white medulla.

Disease Cycle

- ✩ Sclerotia of S. rolfsii are easily spread in soil attached to hand tools, machinery or in splashing water. Long-distance movement occurs by means of sclerotia in plant material or soil. During favorable weather conditions, sclerotia resume activity either by eruptive or hyphal germination. Eruptive germination means that aggregates of white mycelium burst out of the sclerotial rind.
- ✩ An external food source is not required for this type of germination. Sclerotia can germinate eruptively only after being induced by dry conditions or volatile compounds Sclerotia can germinate hyphally more than once. Growth of individual hyphae from sclerotia is in response to availability of exogenous nutrients.

Predisposing Factors

- ✩ The disease is favoured by basal pruning, building of mounds around the stem for irrigation and spreads to the whole plantation through irrigation water apparently because mycelial fragments and sclerotial structures of the fungus remain uniformly distributed in the top soil. S. rolfsiiis able to survive (and thrive) within a wide range of environmental conditions. Growth is possible within a broad pH range, though best on acidic soils.
- ✩ The optimum pH range for mycelial growth is 3.0 to 5.0, and sclerotial germination occurs between 2.0 and 5.0. Germination is inhibited at a pH above 7.0. Maximum mycelial growth occurs between 25 and 35°C with little or none at 10 or 40 °C. Sclerotial formation is also greatest at or near the optimum temperature for mycelial growth.
- ✩ Mycelium is killed at 0°C, but sclerotia can survive at temperatures as low as -10 °C. High moisture is required for optimal growth of the fungus. Sclerotia fail to germinate when the relative humidity is much below saturation.

Control

- Deep plowing (at least 20 cm) with a moldboard extention inverts soil so that organic matter, sclerotia, and plant debris are buried at least 10 cm beneath the surface.
- It is controlled effectively by pruning the stumps at 30 cm and above the ground level because hard cortical tissues around the developed stem prevent entry of the pathogen.
- The compost amendment may increase populations of antagonistic soil microorganisms. Further integrated control of S. rolfisii with nitrogen fertilizers and Trichoderma hazarianum reduces the disease incidence. Further, soil drenching with bavistin (0.05 per cent), brassicol (0.2 per cent) or captan (0.2 per cent) was found effective to control the disease.

4.1.11 Sooty Mould

- Sooty mold is black, superficial growth of fungi especially on the leaf surfaces. The sooty mold results due to interaction of sap feeding insects and non – parasitic fungi. The fungi involved are saprophytic in nature, they do not invade inside the plant tissues, but remain on the surface. The fungi grow on honeydew excreted by sucking insects or on exudates from leaves.
- In mulberry, sooty mould caused by pathogens such as Capnodium sp., Metacapnodium sp. and Euantennaria, have been reported with 90 per cent incidence from Anantapur during 1991. Further sooty mould caused by Chaetothyrium sp. and Curvularia affinis was reported in mulberry during November to February at Central Sericultural Research and Training Institute, Berhampore (Rao *et al.*, 1992). The disease attained in devastating proportion in Malda, Murshidabad and Birbhum districts of West Bengal, India during 1998-1999 and caused panic among sericulture farmers. Further, the sooty mold has been reported from Kashmir.
- The disease is found associated with sap sucking insects mainly white flies *Dialeuropora decempuncta* (Quaintance and Baker) and small-scale insect *Parthenolecanium corni* Bouche (Hemiptera: Coccidae). The disease is very important since it can become epidemic within a short span if no control measures are done. The disease interferes with the photosynthetic activity of leaves and impairs the quality of leaves.

Symptoms

The disease appears in the form of a black powdery mass covering the entire upper surface. Presence of adult white flies or their nymph in the garden which suck the sap from tender leaves resulting in chlorosis and leaf curl. The highly infected leaves fall off from the shoot.

Causal Organisms: *Capnodium* sp., *Metacapnodium* sp. *Chaetothyrium* sp. and *Curvularia affinis*

Predisposing Factors

- ☆ Incidence of whiteflies and other sap sucking insects are the important predisposing factor for the disease incidence. Among the biotic factors, previous 7 days of rainfall and maximum relative humidity are found responsible for the incidence of whitefly incidence.
- ☆ The honeydew excreted by sucking insects serves as a balanced growth medium for fungi. Insect honeydews contain sugars, amino acids, proteins, minerals, and vitamins all are required for growth of the fungus.

Control

- ☆ Management of whitefly is the only way to prevent incidence of sooty mould. Different methods such as biological, mechanical and chemical methods are recommended for the control of white fly.
- ☆ Biological control: Release of 500 pairs of Brumoides suturalis per acresuppresses the population of whitefly population.
- ☆ Mechanical control: Installation of yellow-coloured sticky traps @ 60 traps (2ft x 1 ft) per acre reduces the white fly infestation.
- ☆ Chemical control: Spray of insecticides such as 0.02 per cent Monocrotophos, 0.1 per cent Dimethoate, 0.1 per cent Dichlorvos or 1 per cent Neem oil azadirachtin 1500ppm) is recommended.
- ☆ Safe period: Two weeks (14 days) for both pesticides and Indofil.

4.1.12 Stem Canker/Soft Rot

- ☆ It was reported by Luke and Sathiya (1982) from the mulberry fields in and around Bangalore. Shree and Boraiah (1983) reported 10 to 15 per cent disease incidence in mulberry stem cuttings. Sukumar *et al.* (1991) reported an average mortality of 30 per cent in mulberry nurseries due to the disease.
- ☆ It grows fast on PDA and slowest on mulberry leaf decoction agar. The fungus sporulates heavily in both mulberry leaf decoction agar and PDA.

Causal Agent: *Botryodiplodia theobromae* Pat.(syn: *Lasiodiplodia theobromae*)

(Systematic Position: Kingdom: Fungi, Division: Ascomycota, Class: Sordariomycetes, Order: Diaporthales, Family: Diaporthaceae, Genus: Botryodiplodia, Species: theobromae)

Description of Pathogen

- ☆ Colonies are grayish sepia to mouse grey to black, fluffy with abundant aerial mycelium; reverse fuscous black to black. Conidiophores are hyaline, simple, sometimes septate, rarely branched cylindrical, arising from the inner layers of cells lining the pycnidial cavity.
- ☆ Conidiogenous cells are hyaline, simple, and cylindrical to subobpyriform, holoblastic, annellidic. Conidia are initially unicellular, hyaline, granulose, subovoid to ellipsoide-oblong, thick-walled, base truncate, mature conidia one-septate, cinnamon to fawn, often longitudinally striate, 20-30 x 10-15 μm. Paraphyses when present are hyaline, cylindrical, and sometimes septate, up to 50 μm long.

Predisposing factors: The physical and mechanical injuries are the major predisposing factors of the disease. In case of newly planted mulberry cuttings the cut edges act as portal of entry for the pathogen.

Symptoms

- ☆ The disease develops as a greenish black lesion on the stem and plenty of dark structure producing pycnidial bodies in groups in latter stages.
- ☆ The disease inhibits sprouting of mulberry cuttings (Shree, 1984; Sukumar *et al.,* 1991). Studies on leaf quality revealed that the fungus utilizes D-glucose carbon source best and likes pectin least.

Disease Cycle

B. theobromae Pat. forms chlamydospores and specialized hyphae to overcome unfavorable periods. With favorable conditions, the chlamydospores germinate and the specialized hyphae formed serve as sources of infection to new plants; they infect through wounds or cut edges of the cuttings. Germination tubes infect by means of terminal approsoria.

Control

Cultural control: Deep plough and exposure of soil to sunlight for about a month kills the soil borne pathogens, keep the land leveled to avoid water logging. Keep the land free from weeds during establishment of the saplings/plants.

Chemical control: Prepare 0.1 per cent solution of Indofil M-45 by mixing 1 g in 1 litre of water. Keep the cuttings immersed in Indofil M-45 solution for 30 minutes. Plant these soaked cuttings in nursery beds followed by irrigation after 3-4 hours.

Integrated management: This method involves the dipping of cuttings in Indofil M-45 (0.1 per cent) solution and application of bio-fungicide (like 'Nursery-Guard' containing Trichoderma pseudokoningii). The method of application is as follows.

- ☆ Mix 1 kg Nursery-Guard (*Trichoderma pseudokoningii*) with 60 kg Farm Yard Manure (sufficient for 2000 cuttings). Store the mixture under shade and keep it moistened by adding 10 - 12 litres of water for one week. After one week, broadcast the mixture in to nursery beds @ 2 kg/m2 and mix well in the soil. Soak the mulberry cuttings in Indofil M-45 (0.1 per cent) solution for 30 minutes and plant the cuttings in to treated beds followed by irrigation.

- ☆ For direct plantation of cuttings in to main field, apply 'Nursery Guard' mixture in pits @ 50 g/pit before plantation.

4.1.13 Stem Blight

- ☆ The disease is considered as a minor disease in India.
- ☆ The disease is first reported by Dayakar Yadav and Sukumar (1988) from Tail Nadu during July to October 1985.

Symptoms

- ☆ It attacks tender stems which become pale brown and swollen.
- ☆ The leaves wither and the stem usually breaks at the point of infection. Pycnidia develop in concentric rings inside the necrotic spots. "

Causal Organism: *Phoma exigua*

(Systematic Position : Division: Ascomycota, Order: Dothideales, Family: Pleosporaceae, Genus: Phoma, Species: exigua)

Predisposing Factors: Moderate temperature and high soil moisture conditions favours the disease development.

Disease Cycle

- ☆ The pathogen forms sub-epidermal pycnidia in concentric rings. The pycnidia are globose to sub-globose or obpyriform and ostiolate. Pycnidia wall is pseudoparanchymatous, with outer cell dark and thick walled and inner cells osodiometric and hyaline.
- ☆ Conidiogenous cells are indistinguishable from inner cells. Conidia arise as buds in basipetal succession from the apex of conidiogenous cells.

Control

- ☆ ***Cultural control:*** The affected shoots should be removed and burned. Proper intercultural operations such as deep digging and ploughing to be done.
- ☆ ***Chemical control:*** The cuttings should be soaked in 0.1 per cent Dithane M-45 before planting.
- ☆ ***Biological control:*** Soil application of Nursery guard (A bioformulation made of *Trichoderma pseudokoningii*) reduces the disease severity.

4.2 Bacterial Diseases

4.2.1 Bacterial Leaf Blight

- Bacterial blight caused by (*Pseudomonas syringae* pv. *mori*) is a disease of much economic importance. The disease was recorded on mulberry by Smith (1910), but its occurrence was first reported in India by Sinha and Sexana (1966) from Uttar Pradesh and later by Krishna Prasad and Siddaramaiah (1978) from Kamataka.
- This is prevalent in almost all the mulberry cultivating states of India. In addition of causing heavy defoliation, the disease impairs the nutritive value of mulberry leaves (Sato and Takahashi, 1973).
- Rainy and winter season are most favourable season, disease incidence begins 35 DAP and reaches peak incidence on 70 th days after pruning

Causal organism: *Pseudomonas syringae* pv. *mori*

(Systematic Position: Kingdom: Bacteria, Phylum: Proteobacteria, Class: Gammaproteobacteria, Order: Pseudomonadales, Genus: Pseudomonas, Species: syringae pv. Mori)

Symptoms

- Bacteria produce numerous small, water-soaked, irregular spots on the lower surfaces which also develop on the upper surfaces at later stages. The spot turns blackish brown and becomes slightly angular, surrounded by a yellow margin.
- As the disease advances, spots usually coalesce together and form large irregular dead areas which dry up in dry weather and fall off in wet weather conditions, giving the leaves a distorted and torn appearance. Young leaves are most susceptible to the disease. They appear crinkled, distorted, outwardly curled and ultimately defoliate prematurely.
- Bacteria also infect veins; vein lets and petiole. On the stem, it causes irregular sunken, dark black slits from which bacterial ooze is exuded during the later stages of infection.
- Pathogen leads to both qualitative and quantitative crop loss, it estimated that, 5-10 per cent crop loss annually

Description of the Pathogen

- The pathogenic bacterium is a rod shaped, non-motile, Gram-negative and measures about 0.9 to 1.4 μm x 1.8 to 4.5 μm. It is mainly characterised by small circular shiny white translucent slow-growing colonies on nutrient agar with a smooth and flat surface and an entire edge which gradually becomes undulated and grow slowly.

- ✰ Majority of P. mori isolates grow on a nutrient medium containing saccharose, phenyl red, methyl violet, penicillium potassium ampicillin, sodium acetyl trimethyl ammonium bromide and cadmium chloride. It responds positively to b- oxidative metabolism of glucose and ammonia production but negatively to oxidase test of indole production.
- ✰ It reacts very fast with raffinose while slow with fructose and very slow with glycerol and mannitol.

Disease Cycle

- ✰ Bacteria overwinter in plant parts fallen on the soil. On living plants, they survive epiphytically in buds, on wounds, in exudates or inside the infected tissues.
- ✰ It survives more than 738 days on mulberry leaves packed in paper bags and kept at 5°C, but a little less if kept at room temperature. In general, survival is longer under low temperature and dry conditions than under high temperature and wet conditions.
- ✰ Perennating bacteria are carried to the infection site (stoma, hydathodes, wounds, *etc.*) by rain splash, insects or by wind during wet weather. After landing on wet leaves, they multiply in a film of water or guttation drop and penetrate through stomata, hydathodes or wounds caused by insects or mechanical injuries. Inside the tissues, they multiply and spread intercellular at a rapid rate.
- ✰ Rain by its washing or spattering effect facilitates dissemination to other plants. Necrosis on leaves and twigs appears 20 days after inoculation.

Predisposing Factors: Persistence of high relative humidity during the rainy season and atmospheric temperature fluctuating between 23 to 32°C with little sunshine favours rapid spread and development of the disease.

Resistant Varieties: Bulgaria-24, Bulgaria-3 Thilcseria and Galicene (Coteance, 1974) 'Kariayamagawa' and 'Rogawa'(Takahashi, 1980). M. indica cultivars, *i.e.* Husung China, Shin ichinose, Gumji and Ichihie resistant. (Aktar and Sarvar, 1988) and (Pathan, 1987) Berhampore, Berhampore local, Chanapatna local, Coonoor series C6, CIO, Cll, C14, CIS, C18, CRR, Hakkikalu, Jakkar local, Kajali, MR2, Mysore local, S36, S41, SS4 and YRK Herabam, Italian, Kairyo-neramigheshi, Kosen, Kurangi, M. australis, M. cathagam, M. lambang, M. multicaulis, Okinawa a 1 and 2, Philippine, Popua, Tai-song 2 and 3 and Tsukasguwa (exotic) (Shree and Boraiah, 1988).

Control

- ✰ Foliar application of phytobacteriomycin (0.05 to 0.1 per cent) containing glycerine controls the disease and stimulates the growth as well as the resistance of plants,

- ✰ Streptomycin sulphate, streptocycline and streptopenicillin effective to inhibit the bacteria at 5 to 100ppm on nutrient agar.
- ✰ Fungicides such as Benomyl, Zines sans-317, Epidox nancocare, Dithane and Dithane M-45 highly effective.

4.2.2 Bacterial Leaf Spot

- ✰ The disease is first reported from Korea by Choi *et al.* (1988). In India, Panday and Singh (1989 and Maji *et al.* (1996), from West Bengal.
- ✰ The disease is very severe during rainy season causing huge amount of leaf loss.

Causal Organism: *Xanthomonas campestris* pv. *mori*

(Systematic Position: Kingdom: Bacteria, Phylum: Proteobacteria, Class: Gammaproteobacteria, Order: Xanthomonadales, Family: Xanthomonadacae, Genus: Xanthomonas, Species: campestris pv. mori)

Symptoms

- ✰ Presence of numerous water-soaked spots which initially appear on the lower surface of the leaves. Later tiny black spots appear on the upper surface of the leaves also.
- ✰ The leaves gradually turn brownish, and in the late stage the spots become brownish black surrounded by yellow halo around the spot. The necrotic tissues within the spot fall off forming minute holes, resulting premature leaf fall.

Description of the Pathogen: The pathogen is rod shaped, motile, gram negative. The colonies in nutrient agar are small, circular, convex, slow growing yellow in colour, gelatin and are catalase positive.

Pre disposing factors: Bacterial leaf spot appear in the months of May and continue up to November with maximum severity during July. The BLS disease severity has significant positive correlation with minimum temperatures, maximum and minimum relative humidity, rainfall and number of rainy days and negative correlation with maximum temperature.

Resistant Varieties

Highly resistant: Rotundiloba, MS8, Kolitha-9, kolitha-3, and Mandalaya.

Resistant: S146, Punjab local, Phillipines.

Control: Application of agricultural antibiotics such as Agrimycin 100 ppm (safe period 14 days) or Streptomycin 0.2 per cent or fungicide Dithane M-45 (.2 per cent)- Safe period 14 days.

4.2.3 Bacterial Leaf Scorch

- ✰ Bacterial leaf scorch is caused by *Xylella fastidiosa*, a xylem-inhabiting bacterium that affects water and nutrient transport into leaves and shoots of mulberry and many other shade trees.
- ✰ The disease occurs during summer season.

Causal Organism: *Xylella fastidiosa*

(Systematic Position: Class: Gammaproteobacteria, Order: Xanthomonadales, Family: Xanthomonadaceae, Genus: Xylella, Species: fastidiosa)

Symptoms

- ✰ The symptoms of disease appear initially as marginal leaf desiccation and inward curling. Marginal leaf browning and undulating leaf necrosis with a distinctive chlorotic halo are the symptom in advanced stage.
- ✰ The necrosis initially starts in the desiccated tissues and spreads to the healthy tissues resulting in a scorched appearance. The affected branches show similar symptoms of die back Vigour of infected plants decline followed by premature defoliation.

Description of the Pathogen

- ✰ The bacterium is small (0.25-0.5 x 1.0-5.0 μm), Gram-negative, rod-shaped that lacks flagella for motility and requires oxygen for respiration.
- ✰ The bacterial cells often possess a rippled (undulating) cell wall and terminal fimbriae (surface structures, shorter than flagella, that help to anchor the cells together in the xylem stream).
- ✰ X. fastidiosa, has fastidious nutrient requirements and can be difficult to grow in culture. The bacterium grows slowly on selective medium to form small colonies that appear white to yellow.

Disease Cycle

- ✰ *Xylella fastidiosais* spread primarily by leafhopper insects (Subfamily Cicadellidae) known as sharpshooters, and to a lesser extent, spittlebugs (family Cercopidae). These insects have piercing-sucking mouthparts and subsist on xylem fluid. Both adult and immature (nymph) stages acquire the bacterium when feeding on succulent tissues of infected hosts.
- ✰ As xylem fluid is drawn into the insect, bacterial cells attach to the cibarial pump and the lining of the esophagus (collectively known as the foregut of the insect). There, the bacterium multiplies and forms a biofilm, where it becomes encased and protected in a bacterial "glycocalyx" (made of polysaccharide and protein), extracting nutrients from xylem fluid as it is pumped through the insect. Once an insect acquires the bacterium, transmission to a new host can begin within 1 to 2 hours.

- In the early stages of feeding, bacterial cells become dislodged and are pumped directly into the xylem where systemic movement within the host occurs. An adult can transmit the bacterium for the remainder of its life, whereas nymphs, which shed the foregut during molting, can do so only until the next molt.
- Xylem-feeding insects, particularly leafhoppers, can be polyphagous. Many of the alternative (non-economically important) hosts of X. fastidiosa serve as a food source for potential leafhopper vectors, and many leafhoppers overwinter as adults on these alternative plants. Alternative hosts may be the source of a substantial number of inoculums.
- During the infection process, bacteria are transported from host to host by leaf hoppers, spittle bugs, and sharp shooter insects. Once infected, bacteria move down branches; details of this transport through vascular tissue are not fully understood. Once bacteria reach xylem (vascular) vessels, they multiply and are carried throughout the plant fairly quickly.

Pre-disposing Factors: Occurrence of leaf hopper in the field alternative hosts and temperature more than 28°C influence the development of the disease.

Control

- Detailed studies are not done on the control of the disease. The diseased plants should be removed and destroyed immediately.
- Strict quarantine measures should be followed while transportation of plant materials. Disease-free material should be selected for plantation.

4.3 Viral Diseases

4.3.1 Dwarf Disease

Causal Organism: Mycoplasma

Symptoms

- Appearance of fragile branches noticed on plant
- Yellowing, curling and irregularly arranged mulberry leaves
- Internodal distance become short

4.3.2 Mulberry Ring Spot

Causal Organism: Virus

Symptoms

- Virus transmitted by soil inhabiting nematode
- Infected leaves bear characteristic ring spot and filiform symptoms
- The stunted plant growth can be noticed

4.3.3 Leaf Mosaic Disease

Aphids acts as vector for this transmission of virus which causes leaf mosaic disease

Symptoms

- ☆ Leaves become devoid of chlorophyll, hence mosaic and light greenish patches notice on leaves
- ☆ Plant growth gets stunted, distorted and twined
- ☆ Leaves lose its nutrient status, become undersized

Management

- ☆ Rouguing and burning of infected mulberry leaves and alternate host
- ☆ Using virus free plants for grafting or cutting
- ☆ Washing the hands while handling farm activity with disinfectant (3-10 per cent trisodium phosphate)
- ☆ Evolving virus resistance varieties, adopting micro propagation from meristem or callusulture for obtaining planting material

4.4 Pests of Mulberry

Mulberry is prone to attack by number of pests and they cause abrupt reduction in leaf yield and deteriorate its quality. Feeding such leaves to silkworm results in adverse effect on the cocoon yield and silk quality. Based on the phylum or class, mulberry pests are classified into two major categories called insect and non-insect pests. Insect pests are predominant in mulberry ecosystem whereas only few non-insect species like mites, snails, slugs, nematodes and millipedes are reported to be causing damage to mulberry plants occasionally. The insect pests of mulberry are mainly grouped into sap suckers, defoliators and borers based on their mode of feeding.

4.4.1 Defoliators

1. Leaf Webber, *Diaphania pulverulentalis* Hampson

Taxonomic Position

Phylum: Arthropoda; Class: Insecta; Order: Lepidoptera; Family: Pyralidae; Genus: *Diaphania;* Species: *pulverulentalis*

Description and Life History

The young caterpillar binds the leaflets together with silky secretion, settle inside, and devours the soft tissues of the leaf surface and so the pest is popularly called as leaf webber. During monsoon, the moths emerge from the pupae present in the soil, lay 1 or 2 eggs on the apical portion of each shoot. Total fecundity ranges from 80-150. The eggs are flat, pinkish in color and hatch in 5-7 days. The grown up larva with black head is greenish brown with black markings on the lateral and dorsal regions of body segments. The larval period lasts for 10-15 days after which they pupate in dry leaves or grass at the base of the plant. Pupae are dark brown in color and last for 8-10 days. Adults are greyish white in colour with brown stripes on the forewings and each measures about 10 mm in body length. The life cycle is completed within 29 days.

Alternate Host Plants

Mustard, turnip, radish, red beet, bean, soyabean.

Damage and Symptoms

It is a major defoliator pest known to cause extensive damage to mulberry in Karnataka, Andhra Pradesh and Tamil Nadu. The infestation is observed in mulberry plantations from 12 days through 70 days after pruning. The target area of the leaf webber is the apical portion of the mulberry shoot. The larva binds mulberry leaf blades in tender shoot portion by silken thread hide inside and devours the soft green tissues of the leaf surface. Grown up caterpillars feed voraciously on tender leaves and its faeces can be seen over the leaves below the infested shoot. As this pest devour/damage the apical shoot portion, growth of plants is affected which leads to adverse impact on leaf production.

Mulberry Leaf Roller

Management Practices

- ☆ Clip off infested portion along with the larva into polythene bag and destroy by burning or dipping in 0.5 per cent soap solution (5 g of soap in 1 litre of water).
- ☆ Collect and burn the dry leaves to destroy pupae.
- ☆ Install light traps @ 1-2 trap/acre to attract adult moths and destroy them.

- ✰ Plant dry sticks in all the sides of the garden to attract birds which feed on the larvae.
- ✰ Deep ploughing exposes pupae to sun light and natural enemies.
- ✰ Flood irrigation help to kill pupae.
- ✰ Release *Trichogramma chilonis* egg parasitoids from 5th day after chemical spray @ 1 Trichocard/acre/week for 4 weeks.
- ✰ Release of larval parasitoid *Bracon brevicornis* @ 200 adult wasps.
- ✰ Release pupal parasitoid *Tetrastichus howardii* @ 1 pouch/ac. After the release of these parasitoids, no insecticides should be sprayed in the garden
- ✰ Spray of Chlorfenapyr 10 per cent EC (1.50 ml/L) when the infestation reaches ETL. safe period is 20 days.

2. Bihar Hairy Caterpillar, *Spilosoma obliqua* Walker

Taxonomic Position

Phylum: Arthropoda; Class: Insecta; Order: Lepidoptera; Family: Arctiidae; Genus: Spilosoma; Species : obliqua.

Description and Life History

Bihar hairy caterpillar is a major pest in south India. The female moth lays about 1200 green eggs in clusters on the underside of leaves which hatch in 5-7 days. Body of the newly hatched larvae is dull white in color with small dense hairs and the head is black in color. The anterior and posterior part of the caterpillar is black and the middle part is yellowish brown in color. Fully grown caterpillar attains 5 cm in length. The larval stage takes 27-31 days out of which final instar itself is 5-7 days. Pupation takes place close to mulberry plants in loose soil or leaf litter and it is dark brown in color and measures 2 cm in length. The moth emergence takes place in about 12-14 days. The adult moths are light brown in color with brick red abdomen and having black dots on dorsal side arranged in rows. The wings also possess scattered black dots. The life cycle is completed in about 45-50 days.

Alternate Host Plants

It is a polyphagous pest with a wide host range. Important host plants include amaranthus, cowpea, jatropha, groundnut, jute, sunflower, castor, cotton, green gram, bengal gram, maize, sunnhemp *etc.*

Occurrence

Incidence of Bihar hairy caterpillar,

S. obliqua in mulberry starts with the onset of monsoon. It occurs throughout the year in certain pockets. Peak infestation is seen during March to April and July to November.

Damage and Symptoms

Gregarious young caterpillars feed upon the chlorophyll layer mostly on the under surface and skeletonize the leaves. However, late age caterpillars are voracious feeders, consume entire leaf and cause loss by way of defoliation. In severe cases, only stems are left behind.

Damage Symptoms by Bihar Hairy Caterpillars.

Management Practices

- ☆ Collect the egg masses, caterpillars, affected leaves and destroy them by dipping in 0.5 per cent soap solution or by burning.
- ☆ Follow deep ploughing or digging so as to expose the pupae present in the soil to predatory birds and scorching sun.
- ☆ Flood irrigation also helps in destroying the pupae in the soil.
- ☆ Install light traps to attract and kill the adult moths.
- ☆ Spray of 0.1 per cent Dimethoate 30 per cent EC (safe period 20 days)
- ☆ Release *T. chilonis* @ 4 cards per acre, a week after the spray of insecticide. Do not spray any insecticide after the release of parasitoids.

3. Cutworm, *Spodoptera litura* (Fabricius)

Taxonomic Position

Phylum: Arthropoda; Class: Insecta; Order: Lepidoptera; Family: Noctuidae; Genus: *Spodoptera*; Species: *litura.*

Description and Life History

Adult moth is stout. Fore wings dark with a strongly variegated pattern and paler lines along the veins. Hind wings white with margins having brown colour. The female moths lay eggs in clusters of 200-300 on the underside of leaves and cover them with thick brown scales. Eggs hatch in 4-5 days. The larva is hairless and variable in colour. Final instar caterpillars are 50 mm in length, pale greenish brown in color, stout, cylindrical with transverse and longitudinal grey and yellow bands. The older larvae are night-feeders (nocturnal) and are usually found in the soil around the base of plants during the day. The larval period lasts for 2-3 weeks. They form dark brown pupa in earthen cocoon in the soil which lasts for 2 weeks. The moths are stout, dark with waxy white markings having a brown colour. They are attracted towards light during night. Life cycle takes 30-40 days to complete with 8 generations in a year.

Alternate Host Plants

Cutworm is a polyphagous pest with a wide host range. Important host plants include tobacco, tomato, castor, beet root, carrot, cauliflower, capsicum, potato, radish, cotton, soyabean, cabbage, chickpea, sun flower, mustard, okra, maize, sorghum *etc.*

Occurrence

The cutworm incidence occurs from August to February, mainly in winter season.

Damage and Symptoms

The caterpillars attack shoots of young mulberry plants and cut them, hence the name cutworm. The cut portion of the shoot dries up and falls off. They also feed on mulberry leaves voraciously. In heavily infested mulberry gardens, the plants are seen without branches and sometimes with dried leaves.

Management Practices

- ☆ Collect and destroy egg masses and young caterpillars.
- ☆ Plough infested garden and dig near the base of mulberry plants to expose the pupae present in the soil to sunlight and predators.
- ☆ Install light traps to attract and kill the adult moths.
- ☆ Use Spodolure, a pheromone trap @2 lures/acre twice at an interval of 15 days from 25th day after pruning to attract and kill male moths
- ☆ S. litura is known to be attacked by many natural enemies at various life stages. However release of the egg parasitoid, T. chilonis i s most promising.
- ☆ Spray 0.15 per cent DDVP 76 per cent EC (2 ml/litre) during evening hours, 20 days after pruning. Safe period is 15 days.
- ☆ Spray of 0.1 per cent Dimethoate 30 per cent EC (safe period 20 days)

4. Wingless Grasshopper, *Neorthacris acuticeps nilgriensis* Uvarov

Taxonomic Position

Phylum: Arthropoda; Class: Insecta; Order: Orthoptera; Family: Acrididae; Genus:

Neorthacris; Species: *acuticeps nilgriensis.*

Description and Life History

As the name indicates, the adult grass hoppers are wingless, however, as the hind legs are very strong they are capable of jumping to a long distance. Females are larger than males. Each female lays 6-8 egg pods (each egg pod with 8 to 18 eggs) in soil at a depth of about 3 inches. They hatch in about 28-31 days. The nymphs take 90-95 days to become adults which live for 45-60 days. Thus the total life cycle takes 5-6 months.

Alternate Host Plants

Sunflower, finger millet, groundnut, beans, potato *etc.*

Occurrence

Incidence of this pest coincides with onset of monsoon and continues till post monsoon periods. However, peak infestation occurs during October and declines subsequently with no occurrence from January till onset of monsoon.

Damage and Symptoms

Both nymphs and adults feed voraciously on sprouting buds and leaves of mulberry. Sometimes, they also feed on green bark of affected plants. Branches of plants without leaves are observed in the mulberry garden in case of severe incidence.

Management Practices

- ☆ During early morning hours, they are less active and hence can be collected and destroyed.
- ☆ Deep ploughing immediately after the onset of monsoon to expose egg masses to sunlight and predators.
- ☆ Field sanitation by keeping mulberry garden free from weeds which serves as alternate host plants.

5. May-June Beetle, *Holotrichia serrate* Fabricius

Taxonomic Position

Phylum: Arthropoda; Class: Insecta; Order: Coleoptera; Family: Scarabaeidae; Genus: *Holotrichia*; Species: *serrata.*

Description and Life History

May-June beetles are also known as cock chafer beetles or root grubs or white grubs. The beetles look dark brown in daylight with grey coloured hair all over the body. The abdomen is white. When disturbed and in bright light, they change their colour to light brown and regain their original colour in darkness. The adults are stout and measures 2.5-3 cm in length and 1- 1.5 cm in width. The female lays white colored round eggs individually rolled into earthen balls in the soil at a depth of about 10 cm. Fully grown up grubs are white in color with brownish head and prominent thoracic legs and the length of the grub measures about 56 to 60 mm. The larval development takes about a year with three instars. Pupa is creamy white and semicircular and it lasts for 15 days. After the receipt of first showers they emerge from the soil. The beetles are nocturnal in habit and conceal themselves in the soil during day time.

Alternate Host Plants

Neem, ber, grapes, guava, moringa, mango, *Melia azadirachta,* rose, *Acacia,* Jamun, fig, *Ficus religiosa, Ailanthus,* tamarind, *Cassia fistula,* groundnut, maize, jowar, potato, bajra, sweet potato, tapioca, chilly, sugarcane, tobacco, coffee *etc.*

Occurrence

Their infestation coincides with the onset of monsoon and an occasional pest to mulberry crop in south India.

Damage and symptoms

The early hatched grubs initially feed on organic matter and later start feeding on roots or rootlets thereby causing damage to the host plants. The adult beetles are observed feeding on the foliage. During night time, adult beetles enter into mulberry garden in swarms and feed voraciously on the foliage, leaving only the stem portion. It appears as cow grazing and lot of small black faecal pellets appear below the plants.

Management Practices

- ☆ After first monsoon keep a vigil for adult beetles in the mulberry garden. Collect them and destroy by putting in kerosene solution.
- ☆ Installation of light trap and collection of adult beetles during night in kerosene mixed water.
- ☆ Tying up of few fresh neem branches (with leaves) to mulberry plants in different parts of the garden to attract the adults followed by their collection and destruction.
- ☆ Ploughing just before monsoon helps in exposure of various stages of the pest to natural enemies.
- ☆ Drenchthe soil with 0. 2 per cent Chlorpyriphos 20 per cent EC to kill the grubs.

4.4.2 Borers

1. Mango Stem Borer, *Batocera rufomaculata* DeGeer

Taxonomic Position

Phylum: Arthropoda; Class: Insecta; Order: Coleoptera; Family: Cerambycidae; Genus: *Batocera;* Species: *rufomaculata.*

Description and Life History

The adults are 3-5 cm in length, dark with a fine greyish vestiture, pronotum with 2 kidney shaped orange yellow spots, the basal third of the elytra with numerous black tubercles and several yellowish spots that are variable in number and shape. Fully grown grubs are cream colored with a dark brown head and are up to 10 cm long. The female cuts the tree bark and lays eggs singly into these cuts and cover with a viscous fluid. Fecundity ranges from 150-200. Eggs are shiny white in colour, oval shaped and measuring 5-7 mm long.

Egg hatches within 10-15 days. On hatching the neonate grubs initially feed under the bark and then start to tunnel into the sapwood of the trunk. The grub lives for long time of about one year and hibernate in winter inside the dry shoot. When climate warms up they become active and pupate inside the stem, then adult emerge in May– June by a short tunnel running to the exterior and making a circular exit hole and start egg laying from July to August. The maximum life recorded for the adult is eight months. The adults can travel considerable distance and have good dispersal ability. A generation takes more than one year.

Alternate Host Plants

Mango stem borer is a polyphagous pest distributed worldwide and infesting number of horticultural crops and forest trees including mango fig, guava, jack fruit, pomegranate, apple, avocado, cashew nut, durian, eucalyptus, rubber *etc.*

Occurrence

Damage of this borer on mulberry is noticed throughout the year.

Damage and Symptoms

Once the grub enters into the shoot, it starts feeding voraciously and creates tunnel inside the stem by boring towards upward direction. The grubs, because of their size make large tunnels of about 2-3 centimeters which interfere with sap flow and affect foliage yield at initial stages, then wilting of shoots appears and finally the plant dries. The symptoms of *B. rufomaculata* infestation in mulberry confuses with the symptoms of root rot. However, the frass ejected from the grub boring is occasionally visible in few plants. The grub requires stem girth of at least 6-7cm diameter to successfully develop into an adult. Therefore, it is much prevalent in older mulberry garden than younger ones which have thin stumps. In severely affected garden the exit holes of adults are often visible on the stem just above the ground level.

Management Practices

The mango stem borer feeds internally and thus difficult to identify their presence as well as to control, once it enter the mulberry stem/shoot. It is also because that dry shoots does not carry the pesticides to reach the grub. However following management practices need to be adopted to check the pest.

- ☆ Keep the mulberry garden clean and follow recommended package of practices.
- ☆ Exclude alternate host plants in the vicinity of mulberry garden.
- ☆ Monitor the presence of adult beetles in mulberry garden during summer and monsoon seasons. Collection and destruction of the beetles before egg laying helps to avoid infestation of grubs.
- ☆ Set light traps to attract the adult beetles
- ☆ Collect and destroy the grubs which are exposed out at the time of weeding or ploughing.
- ☆ Mechanically remove the grubs from the infested trunk holes by using iron wire or hook.
- ☆ Prune and destroy the affected branches and paste the cut ends with 5 per cent Copper Oxychloride (50 g/litre).
- ☆ Two sprays on the trunk portion with 0.04 per cent Chlorpyriphos 20 per cent EC (2ml/litre) at fortnightly intervals with onset of monsoon in summer (May-June) especially to the mulberry gardens in the vicinity of mango orchard.

2. Stem Girdler, *Sthenias grisator* (Fabricius)

Taxonomic Position

Phylum: Arthropoda; Class: Insecta; Order: Coleoptera; Family: Cerambycidae; Genus: Sthenias; Species: grisator.

Description and Life History

The adult beetles are stout with long antennae and well developed mandibles. Female girdles the bark of the stem at night and deposit eggs beneath it. The eggs hatch in a week and the grubs tunnel into the branches and feed. They pupate inside the tunnel and the life cycle is completed in about 7-8 months.

Alternate Host Plants

Major alternate host plants are casuarina, mango, jack, crotons, grapes *etc.*

Occurrence

It is noticed throughout the year.

Damage Symptoms

The adults are stout built with long antennae and well developed mouth parts. Female deposits eggs underneath the bark of the girdle with the mouth parts. The grubs tunnel in to the wilting branches and feed. The beetle has a peculiar habit of ringing the stem. The bark and wood are neatly cut around the stem and branches leaving a clean girdle. The portion above the girdle gradually wilts and dies.

Management

Cutting and burning the branches and stems showing the symptoms of beetle attack.

Removal of caterpillars by use of hooks and swabing the base of main stem or branches with 0.1 per cent Malathion emulsion.

3. Stem Borers *Indarbela quadrinotata* (Lepidoptera: Metarbelidae)

The bark caterpillar infestation can be easily detected by the presence of winding galleries of flossy web on the stem and near forks or angles of branches. The caterpillar remains in the tunnel at the axils of branches and moves about conceal inside the silken gallery and feed on the bark by scrapping. Heavy infestation retards the growth and affects the leaf yield.

Management

Removal of larvae from galleries on the stem using hooks and swabing with 0.1 per cent Malathion.

4.4.3 Sap Suckers

1. Pink Mealybug, *Maconellicoccus hirsutus* (Green); *Paracoccus marginatus* Williams and Granara de Willink

Taxonomic Position

Phylum: Arthropoda; Class: Insecta; Order: Hemiptera; Family: Pseudococcidae; Genus: *Maconellicoccus;* Species: *hirsutus.*

Description and Life History

Adults of both female and male are about 3 mm long. Females are pink in color with a white waxy covering and have no wings. Males have a pair of wings and two long waxy tails and are capable of flight. Female can reproduce parthenogenetically and each female lays approximately 350-600 eggs in an ovisac covered with cotton like mealy substance. The eggs will hatch in 6-9 days. Males have four nymphal instars while females have only three. Nymphal period lasts for 23-27 days and total lifecycle completes in a month period. There will be 10-12 generations in a year.

Alternate Host Plants

They are highly polyphagous and so far more than 350 host plants have been recorded in the world. Important host plants are hibiscus, beans, pumpkin, croton, chrysanthemum, citrus, grapevine, guava, coffee, sugarcane, soybean, mango, pigeon pea, maize, cotton, teak *etc.*

Occurrence

They occur on mulberry throughout the year, but the incidence is high in summer months (March to August). Their population is negligible during rainy season.

Damage and Symptoms

The nymphs feed by sucking the sap from tender leaves and stem portion. Hence the affected apical shoots show bunchy appearance due to curling of leaves, shortening of internodes and thickening of stem. This symptom is popularly known as 'Tukra' in India. In advance stages of infestation black sooty mould is developed in the affected area due to growth of fungus on the honeydew secreted by the mealybug. The pest makes both quantitative and qualitative loss in mulberry production due to retarded growth of plants and depletion in nutritional value of the leaves.

Management Practices

- ✰ Clip off infested apical shoots and destroy by burning or dipping in soap solution.
- ✰ Do not grow alternate host plants of the mealybug in the vicinity of mulberry gardens.
- ✰ Spray 0.05 per cent Dimethoate (36 per cent EC) 12-15 days after pruning. Safe period to silkworm is 20-25 days.
- ✰ Release predatory ladybird beetles *Cryptolaemus montrouzieri* @ 250 adults or *Scymnus coccivora* @ 500 adults/acre/year in two split doses at an interval of six months.

2. Mulberry Thrips, *Pseudodendrothrips mori* (Niwa)

Taxonomic Position

Phylum: Arthropoda; Class: Insecta; Order: Thysanoptera; Family: Thripidae; Genus: *Pseudodendrothrips*; Species: *mori*.

Description and Life History

Mulberry thrips reproduce both sexually and parthenogenetically. Each female lays about 30-50 bean shaped whitish eggs measuring 0.25 mm in length and 0.10 mm in breadth on ventral surface of tender leaves which hatch in 6-8 days. The newly hatched first instar nymph is initially colourless and transparent with a pair of dark red compound eyes and gradually changes to light creamy yellow. There are two nymphal stages with 6-7 days. Second instar nymph measures about 0.70 mm length and 0.23 mm breadth. After nymphal stages, it enters into soil and become inactive and non-feeding stages called pre-pupa and pupa. The pupa is yellow coloured, characterized by two pairs of short wing pads. Adults are about mm long. Female is yellow in colour, while male is darker. Total lifecycle is completed in 20 - 22 days with 15 generations in a year.

Alternate Host Plants

Very few plants l ike *Ficus, Tridax procumbens, Camellia sinensis, Leersia hexandra* are reported as alternate hosts of *P. mori*.

Occurrence

Occurs throughout the year, but incidence is severe in summer months (April – May) and least in rainy season (October- November).

Damage and Symptoms

Nymphs and adults are found mainly on the underside of the leaf. They pierce the epidermis of mulberry leaves using their lacerating mouthparts and extract the plant sap. During laceration, they secrete saliva which coagulates the sap resulting in the formation of white streaks in the early stage followed by silvery blotches which are mixed with small black spots of thrips faeces. This symptom helps in distinguishing damage by *P. mori* from that caused by other mulberry pests. As the leaf tissue dries beneath the epidermis the silvery patches turn brown and become depressed. In acute cases serious drying of the leaf tissues results in leaf curl and these leaves shrink harden and ultimately fall. Stunting, leaf curling and deformation are also observed in severely affected gardens. Feeding thrips affected mulberry leaves to silkworm results in adverse impact on economic traits and cocoon yield.

Management Practices

- ✰ Mulberry field should be thoroughly cleaned after harvest by removing small side branches, dead leaves and weeds in order to eliminate any developmental stages of thrips on them.
- ✰ Periodical ploughing and digging of mulberry field helps in exposing the thrips pupae to hot sun and natural enemies.
- ✰ Water jetting or sprinkler irrigation is effective in reducing thrips population.
- ✰ Providing frequent irrigation helps in increasing the pupal mortality in soil thereby reducing the thrips emergence.
- ✰ Spray of 0.1 per cent Dimethoate 30 per cent EC (3ml/litre) 15 days after pruning. Safe period is 20-25 days.
- ✰ Release of *S. coccivora* @ 500 adults or *Chrysoperla* @ 1000 eggs/acre, a week after the insecticide spray.

5. Spiralling Whitefly, *Aleurodicus dispersus* Russell.

Taxonomic Position

Phylum: Arthropoda; Class: Insecta; Order: Homoptera; Family: Aleyrodidae; Genus: Aleurodicus; Species: dispersus.

Description and Life History

Each female lays about 40-70 pale yellow coloured eggs with a short pedicel on the underside of leaf. The eggs are laid along with deposit of waxy secretion in a characteristic spiraling pattern, hence the name 'spiralling whitefly'. Incubation period is 4-6 days. There are four nymphal stages. The first instar nymphs are mobile (crawlers) and suck the sap from the tender leaves. Subsequent instars are immobile which are found on the underside of older leaves. They produce whitish waxy filaments and copious quantities of honey dew.

The final instar is the pupa from which adults emerge. The nymphal period exist for 14-20 days. The adult flies are 2 mm in length with white powdery wax material over the wings and whole body. Life cycle is completed in a month and has 10-12 generations per year.

Alternate Host Plants

Mango, custard apple, apple, papaya, banana, sweet potato, cassava, avocado, guava, citrus, capsicum, brinjal, tomato, pepper, rose, hibiscus, coconut, *etc.*

Occurrence

It is found to cause damage throughout the year with peak incidence during summer.

Damage and Symptoms

Spiralling whitefly remains on ventral surface of leaves in colonies and cause direct feeding damage by sucking plant sap and indirect damage due to the honeydew and white waxy material produced by the insect. Adults and nymphs feed on plants by sucking plant sap from the leaf through a slender stylet which results in leaf curling, chlorosis, defoliation and stunted growth. Large amount of powdery wax material secreted by all the stages of pest is readily spread through wind, causing nuisance. The honeydew excreted by these insects will fall on the upper surface of the lower leaves which becomes a medium for developing sooty mould fungus, *Capnodium* sp. This in turn, interferes with photosynthetic process by not allowing enough light to reach the cytochrome tissues of the leaves.

The sooty mould may also increase thermal absorption and raise leaf temperature, thus causing reduced leaf efficiency and further deterioration in the nutritional quality. Feeding spiralling whitefly infested mulberry leaves to silkworm will affect its growth and development which leads to adverse impact on cocoon production.

Management Practices

- ✰ Remove and destroy the infested leaves and install yellow sticky traps @ 75 per acre.
- ✰ Spray of a strong jet of water in the affected mulberry garden will help to reduce the pest population below economic injury level without applying any chemical insecticides.
- ✰ Spray 0.05 per cent Dimethoate 30 per cent EC (1.5 ml/litre) about 15 days after pruning and second spray with 0.15 per cent DDVP 76 per cent EC (2 ml/litre), one week after first spray (safe period 20-25 days).
- ✰ Release *S. coccivora* @ 500 adults or *Chrysoperla* @ 1000 eggs/acre, a week after the insecticides spray.
- ✰ Two parasitoids namely *Encarsia quadeloupae* and *Encarsia haitiensis* are known to parasitise *A. dispersus* in India.
- ✰ An egg predator *Axinoscymnus puttarudriahi* was identified and it could also be used to destroy the pest in egg stage itself.

6. Jassid, *Empoasca flavescens* Fabricius

Taxonomic Position

Phylum: Arthropoda; Class: Insecta; Order: Hemiptera; Family: Cicadellidae; Genus: Empoasca; Species: flavescens.

Description and Life History

Adults are pale green or greenish yellow in color and about 3-4 mm in length. Body is wedge-shaped with whitish markings on the head and thorax. The head is prolongated forward as a smooth, flat, triangular structure with a pair of antennae possessing sensoria. The thorax is simple and abdomen is tapering at the posterior end. The hind legs have two parallel rows of spines which extend all along the hind tibiae. The wings are held like a roof over the abdomen. The nymphs resemble adults, but they lack wings until the fourth instar. Adults and nymphs move sideways. Female lays 20-30 eggs on the lower surface of the leaves below the epidermis. Eggs are pale yellow in color and elongated in shape. The egg hatches into nymph within 6-13 days. The nymph undergoes five instars and becomes adult in 8 to 22 days. Nymphal period is 19-27 days. Pupation takes place on the leaf itself. There are 10-12 generations in a year.

Alternate Host Plants

Castor, tea, okra, cotton, beans, brinjal, potato *etc.*, are prone to severe attack of jassid.

Occurrence

It is considered to be a minor or occasionally serious pest of mulberry. The insect remains active throughout the year, but maximum population build up occurs during November to January, that is mostly in the winter season.

Damage and Symptoms

Both nymphs and adults damage the plant by sucking the sap of young leaves and tender shoots. At the time of sucking cell sap inject toxins in to the plant tissues. The early symptoms are the appearance of yellow or brown patches at the margin of the leaves followed by distortion of leaf veins. Finally, leaves curl upward becoming cup shaped, margins turn brown, dry and wither off prematurely. This characteristic symptom is known as "Hopper burn".

Management Practices

- ☆ Set up light traps and yellow sticky traps to destroy adult population.
- ☆ Sprinkler irrigation is effective in controlling the pest.
- ☆ Spray neem oil (3 per cent) with fish oil rosin soap (2 per cent). Safe period is 10-12 days.
- ☆ Spray of a strong jet of water in the affected mulberry garden help to reduce the pest population below the economic injury level.
- ☆ Cultivation of cluster bean, cowpea, black gram or groundnut as intercrops in mulberry encourage to buildup natural enemies like coccinellids and spiders.
- ☆ Spray 0.1 per cent Dimethoate 30 per cent EC (3ml/litre). Safe period is 25 days.

4.4.4 Soil Inhabiting Insect Pests

1. Termites, *Odontotermes* sp.

Taxonomic Position

Phylum: Arthropoda; Class: Insecta; Order: Isoptera; Family: Termitidae; Genus: *Odontotermes.*

Description and Life History

Termites or white ants are soil inhabiting social insects. They are found in all types of soils but frequent in sandy and red loamy soils. They live in colonies with different casts such as workers, soldiers and queen. The queen has a very large abdomen and lays about 75,000 eggs in a day. Eggs hatch in 24 to 90 days. Duration of development of young ones vary with casts and environmental conditions. The reproductive casts mature in 1-2 years and there will be only one queen per colony which lives for 5-10 years.

The queen is fed by the workers and is confined to royal chamber with king. The king's life is very shorter than that of queen. When the king dies, a new one replaces it. The workers usually develop from fertilized eggs but remain stunted as they feed on ordinary food. The workers perform all the works except reproduction and defence. The soldiers develop from unfertilized eggs and remain under developed. They defend the colony against enemies such as ants and other predators.

Alternate Host Plants

Termites are highly polyphagous insects attacking all most all the agricultural, horticultural and forest tree plants.

Occurrence

They occur when rain recede or from October onwards and continues till the onset of monsoon rains.

Damage and Symptoms

Termite damage is mainly observed in rainfed gardens. In mulberry nursery and new plantation, they attack below ground portion. They feed on the bark and hard wood. Hence, cuttings dry up and no sprouting takes place. In old plantations, they first infest the dry twigs. Later they slowly move to live twigs. They form foraging galleries inside the main stem and extending below the ground. In case of pruned plants they form a sheath around the twigs and feed on them. Thus they affect the sprouting buds.

Management Practices

- ☆ Remove the dead and dried twigs and leaves.
- ☆ Flood irrigation help in keeping termites away.
- ☆ Locate the termite mounds if any nearby mulberry gardens and destroy by breaking mounds and kill the queen. When once the queen is killed or destroyed the colony gets abandoned by them.
- ☆ Prepare a solution of Chlorpyriphos 20 per cent EC @ 3ml/litre and pour into the mound followed by closing the mound hole with wet earth.
- ☆ In established plantation, soil drenching with 0.1 per cent Chlorpyriphos 20 per cent EC to be practiced.
- ☆ Treat the mulberry cuttings with 0.1 per cent Chlorpyriphos 20 per cent EC solution before planting.

4.4.5 Non Insect Pets

1. Giant African Snail, *Achatina fulica* Bowdich

Taxonomic Position

Phylum: Mollusca; Class: Gastropoda; Order: Stylommatophora; Family: Achatinidae; Genus: *Achatina;* Species: *fulica.*

Description and Life History

Giant African snails are large in size (long, narrow and conical in shape, reaching a length of 20 cm), with a light brown shell to protect the body, having brown and cream bands alternatively lined. The shell size attains up to 5-10 cm. They are nocturnal in habit and hide beneath fallen leaves, inside peeled off mulberry bark or under stones during day time and come out from hiding places during night to attack host plants and cling to the dorsal and protected surfaces of leaves. Though bisexual in nature, self reproduction is not possible. They are active during rainy season. They start laying eggs beneath the soil, dried leaves or any debris 2-3 weeks after mating. Eggs are round, yellowish in color, surrounded by mucus substance and are laid in groups of 30 to 300 eggs. They lay about 1000 eggs during their life time of about 3 to 5 years with 2-13 egg clutches. The young ones attain maturity in about 9 weeks. They undergo hibernation and aestivation during adverse climatic conditions.

Alternate Host Plants

Snails are highly polyphagous and nominated as one among 100 'world's worst' invaders. They have a wide host range and infest crops such as coffee, mango, papaya, rubber, teak, tea, coconut, sunflower, maize, gram, onion, garlic, peas, beans, bell pepper, brinjal, pumpkin, cucumber, cabbage, cauliflower, bhendi, banana, sweet lime, marigold *etc.*

Occurrence

Though snail activity is noticed from August to January (mostly during rainy season), their incidence is severe from October to December. Their occurrence is also noticed in climatic conditions with high humidity (> 80 per cent) and moderate temperature (9 - 29°C) which are more congenial for the population build up.

Damage and Symptoms

Giant African snails feed on tender leaves, tender bark and stem and cause damage. The infested leaves show circular holes in the centre. Such damage results in stunted growth of mulberry besides leaf yield loss. It is also established that when mulberry leaves from snail infested gardens are provided to silkworms, they consume very less leaves due to presence of mucus like substance on the leaves secreted by the pest which affects cocoon production.

Management Practices

- ✰ Hand pick different stages of snails and destroy by burning or immersing in 25 per cent salt solution (1 kg salt in 4 litre of water).
- ✰ During rainy season, in the evening hours heaps of wet gunny sacks or papaya stem waste may be placed near the hide outs in the mulberry garden. Next day morning the snails hiding below these may be collected and destroyed by immersing in salt solution.

- ☆ Regularly remove the debris and unnecessary waste materials from the garden and maintain cleanliness to avoid snail population build up.
- ☆ Deep ploughing helps in exposing snails and their egg masses present in the soil to their natural enemies.
- ☆ Snails do not move on dry areas. Hence areas surrounding their hiding places may be kept dry by spreading materials like saw dust, ash *etc.*
- ☆ Use bran bait (Rice bran, jaggery and Methomyl @ 60:6:1 ratio) to attract and kill the snails.
- ☆ Snail-kill (2.5 per cent Metaldehyde) pellets may be spread in mulberry garden in alternate rows (2 to 3 pellets per spot) during evening hours in rainy season or after irrigation. Next day collect the dead snails available near these pellets and destroy. This process may be continued for a week or so.
- ☆ Metaldehyde pellets may also be placed near their hiding places, compost pits, dumping yards for suppressing snail population. About 2 kg Metaldehyde pellets are required per acre of mulberry garden. Metaldehyde is found to be non-toxic to silkworms.

2. Black Slug, *Laevicaulis alte* (Ferussac)

Taxonomic Position

Phylum: Mollusca; Class: Gastropoda; Order: Systellommatophora, Family: Veronicellidae; Genus: *Laevicaulis;* Species: *alte.*

Description and Life History

Slugs are soft bodied, dorsally black and ventrally brown, 4 cm long. Its skin is slightly tuberculated. The central keel is beige in colour. This slug has a unique, very narrow foot, juveniles have 1 mm wide foot and adults have foot of 4 or 5 mm wide. The tentacles are small 2 or 3 mm long and they are only rarely extended beyond the edge of the mantle. Eggs are laid in groups (38 to 80 eggs) and arranged in the form of bead. Freshly hatched young ones feed on the leaves by making small holes. It leaves behind a salivary, shining and slimy material wherever it crawls.

Alternate Host Plants

Coffee, tea, cotton, rice, beans, tomato, cucumber, apple, spinach, carrot *etc.*

Occurrence

The slugs occur mostly during rainy and winter seasons and also common in mulberry gardens situated nearby canals and swamps.

Damage and Symptoms

Presence of mucus on plant parts affects the quality of foliage. Presence of shot holes on leaves caused due to feeding by slugs. Drying of shoots due to scraping of bark of the stems.

Management Practices

Soil application of l ime powder (20kg/ac) dehydrates the slug and brings about mortality.

3. Red Spider Mite, *Tetranychus ludeni* (Zacker)

Taxonomic Position

Phylum : Arthropoda; Class: Prostigmatta; Order: Acarina; Family: Tetranychidae; Genus: Tetranychus; Species: ludeni.

Description and Life History

So far 15 species of mites are found to infest mulberry out of which Tetranychus ludeni, T. equitorius, T. telarius and Aceria mori are the important ones. They are also known as spider mites and are found on almost all the plant parts like leaves, buds, scales, nodes and apical shoots. Adult female lays about 30-70 spherical red colored eggs on the lower surface of the leaves which hatch in 2 to 6 days. The nymphs are pinkish, greenish or orange in color depending on the species whose development takes about 10-12 days. Adults live for 7 to 12 days.

Alternate Host Plants

Brinjal, cotton, okra, Amaranthus, beans, citrus, water hyacinth, chrysanthemum.

Occurrence

Although mites occur throughout the year, their incidence is severe during summer months.

Damage and Symptoms

Both nymphs and adults cause similar type of damage. They suck the plant sap by piercing their stylet mouth parts into the tissue. The affected plant part turns greyish white and finally withers. Infested leaves also show white speckles at the place of feeding which later turns into large patches. Infested plants are often covered with extensive fine webbing. The undersurface of infested leaves show silken thread spun across them under which they crawl and also lay eggs.

Management Practices

Clip off affected portion and destroy by burning.

Use sprinkler irrigation to dislodge the population.

Spray 0.5 per cent Formathion (safe period 25 days).

3. Root-knot Nematode, *Meloidogyne incognita* (Kofoid and White) Chitwood

Taxonomic Position

Phylum: Nematoda; Class: Secernentea; Order : Tylenchida; Family: Meloidogynidae; Genus: *Meloidogyne;* Species: *incognita*.

Description and Life History

M. incognita overwinters in the form of egg in the affected roots and soil. It can survive in the soil without host plant for about 6-12 months. The life cycle is completed passing through three stages *viz.*, egg, larva and adult. Egg and larva are the main sources of infection. Eggs are ellipsoidal and larvae are vermiform. Larvae invade the roots through root tips by making the hole with the help of their stylets. They start feeding on the parenchymatous cells of the root. While feeding larvae inject the salivary secretion containing proteolytic enzyme that leads to the enlargement of cells resulting in galls/knots in the roots.

The larvae undergo three moults and mature in to adults. The male is cylindrical worm like transparent with stripes across the body while the female is pear shaped with elongated pyriform body. Reproduction is mostly by partheno- genesis. Each female lay 300-500 eggs covered with gelatinous substance. It has six generations in a year and one life span is completed within 30-40 days.

Alternate Host Plants

It has a wide range of host plants and attacks more than 2000 species of plants including almost all agricultural, horticultural, oilseeds, ornamental, plantation and other crops.

Occurrence

Widespread throughout the year and more prevalent in sandy soils under irrigated conditions.

Damage and Symptom

The incidence and intensity of this nematode is reported to be very high in red sandy soil followed by red loamy soil, while in black cotton soil the incidence is found to be very poor. Considering the farming practices, high incidence and intensity are found under irrigated mulberry gardens while poor or negligible incidence are noticed under rainfed conditions. The magnitude of crop loss depends up on the population density of the nematode present in the soil. It also reduces the nutritive value of leaves particularly protein content which is an important nutrient for silkworm growth.

Management Practices

- ☆ Use nematode free saplings for new plantation
- ☆ Deep digging and soil solarization during summer exposes the nematode eggs and larvae leading to desiccation.
- ☆ Soil amendment with oil cakes like Neem/pongamia @ 2 MT/ha year in 4 equal split doses at 3 months interval
- ☆ Intercropping with plants like marigold, sesame or sunhemp reduces root galls and egg masses of nematode

- ✰ Apply Bionema (*Verticillium chlamydosporium*) mixture @ 200 g/plant around the exposed roots of affected plant (Bionema mixed with neem oil cake and FYM in the ratio of 1:24:200) for 3 times in a year at an interval of 4 months during cultural operations.
- ✰ Apply Nemahari, a plant based eco friendly formulation @ 40 kg/ha (mix 40 kg Nemahari with 400 kg FYM) in 3 split application on 70 and 150 days after first application.

Severely affected mulberry plants show stunted growth with marginal chlorosis and necrosis of leaves. Knots/galls formed on the roots are spherical with varying size. Young galls are very small and yellowish white while old ones are big and blackish brown in colour. Due to this infection vascular tissues and cortex of roots are disorganized, hampering the translocation of water and minerals from soil to aerial parts of the plants.

Chapter 5

Propagation Methods in Mulberry

5.1 Introduction

Today mulberry is cultivated in about 2.90 lakh hectares in India and the stress is upon productivity. Earlier the mulberry production per hectare was around an average of 12,000 to 15,000 kgs of leaf under irrigated conditions. Presently the average production has been tremendously increased due to the adoption of new technology like proper maintenance of mulberry garden and planting high yielding varieties while the average mulberry production per hectare has enhanced from 25,000 kg to 35,000 kgs.

5.2 Propagation of Mulberry

Mulberry can be propagated in two ways sexual and asexual.

5.2.1 Sexual Propagation

In mulberry the sexual propagation is through seedlings, particularly the seed propagation carries a varied population, this to utilize in selection and hybridization. For seed germination certain prerequisites are needed to be fulfilled such as selection of quality seed, preparation of land, and the seed should be selected such that can definitely germinate. This is possible only when the seed is subjected to suitable environmental conditions, embryo of seed is alive, and healthy, in internal conditions of seed are favourable for germination. The fresh seeds will have greater germination rate than the stored once. Seeds must be washed with fresh water until the flesh of fruit is withdrawn and dried well, however minimum moisture percentage should be maintained, *i.e.*, at least 6 per cent. Sowing of seeds may be by way of broadcasting or sowing in lines.

5.2.2 Asexual Propagation

In asexual type of propagation vegetative plant parts are used. In mulberry the propagation is mainly of three types.

- ✰ Propagation by Cutting.
- ✰ Propagation by Grafting.
- ✰ Propagation by Budding.

In case of mulberry, mostly through "Cuttings" only the propagation is practiced. By asexual method of propagation the inherited characteristics of parent stock can be retained. The desired characteristics can be carried to next generation. The asexual propagation in mulberry is carried out by grafting, cutting, layering.

5.2.2.1 Stem Cutting

- ✰ The ideal mulberry variety has high rooting ability. Therefore, it is easier to multiply these varieties through stem cuttings. The process starts with harvesting of shoots from a mature mulberry garden.
- ✰ Precautions are taken to keep the garden free from disease and pests. It keeps in mind that stem cuttings should be selected from at least 6 – 8 months old shoots. Cuttings of 15-18 cm length and 10mm diameter with 3-4 healthy buds are prepared using sharp billhook or secateur without any damage to the bark. While preparing the cuttings, upper immature and lower thick portion of the shoots are rejected.
- ✰ May also take precautions to see that drying of the cuttings can be avoided by putting the cuttings in shade covered with wet gunny cloth.

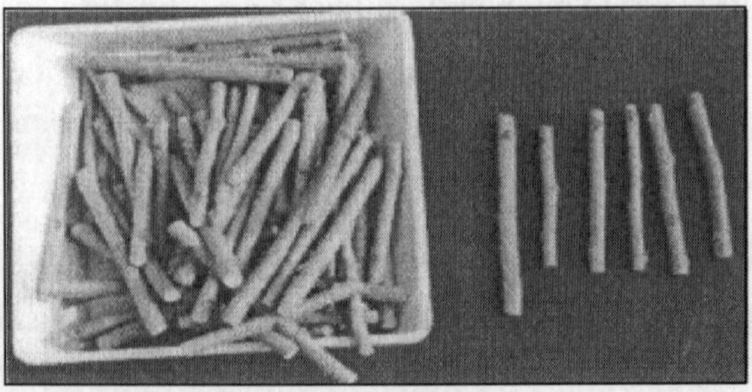

Stem Cuttings.

- ✰ They will not develop roots. Before planting cuttings in the nursery, the cuttings are treated with 0.1 per cent. Dithane M-45 solution for 20-30 minutes for avoiding fungal disease.

- ☆ Suitable temperature can help in multiplication. Hence, temperature of 20-25°C is considered ideal for cutting propagation. Soil temperature is also important. High humidity can be an added advantage in the process of multiplication. Perhaps, it is simple to understand that high humidity will help in avoiding drying of cuttings in nursery.
- ☆ Now, you shall try to learn how the cuttings are planted in the nursery beds in a slanting position exposing only one bud above the soil. Also know the space that is to be maintained while planting cuttings. The distance between rows should be 20 cm and between cuttings 10 cm. Soil should be pressed around the cuttings firmly to ensure contact of cuttings with the soil to develop roots.
- ☆ Figures given shows how the cuttings can be planted at a spacing of 10/20 cm. During the process of development of roots, the cutting should not be disturbed. Weeds also hamper the growth of cuttings. So, weeds can be removed 45 days after planting cuttings with the help of a sharp weeding sickle without disturbing the cuttings.

After one month of planting, urea at the rate of 10 g/sq meter is applied to the nursery and irrigated immediately. Subsequently, the irrigation is done once in every 4 days to get uniform size saplings with good vigour. The saplings are transplanted after 4 months.

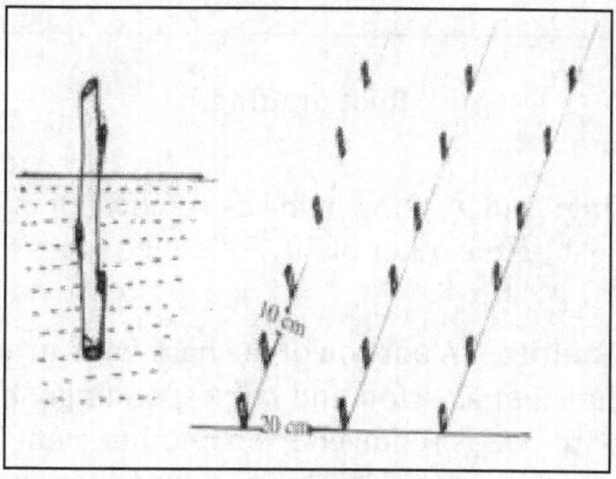

Plantation of Cuttings in Nursery.

5.2.2.2 Grafting and Budding

Depending upon the part used as the stock, the grafting can be classified into root, shoot and bud grafting.

- ☆ ***Root Grafting:*** As mentioned earlier, there are various methods of vegetative propagation. Root grafting is one of the methods. It is practised with mulberry varieties, which cannot be propagated through stem cuttings. The procedure is usually adopted with temperate mulberry varieties. The procedure involves inserting a small branch of a plant into a rooted plant. The branch that is inserted is called as scion and that into which the scion is inserted is known as stock. It is important to choose the right variety of mulberry as scion. The stocks can be raised by sowing the seeds uniformly in a nursery bed. The seeds should be covered with soil to avoid drying. The seeds are sown and allowed to grow with 3 cm distance in between. This can be done by thinning the seedlings. Besides this, water is sprinkled to help in germination. The seedlings are used for grafting. However, it can be indicated here that root grafting is expensive but unavoidable in mulberry propagation, where response for multiplication through stem cuttings is poor.

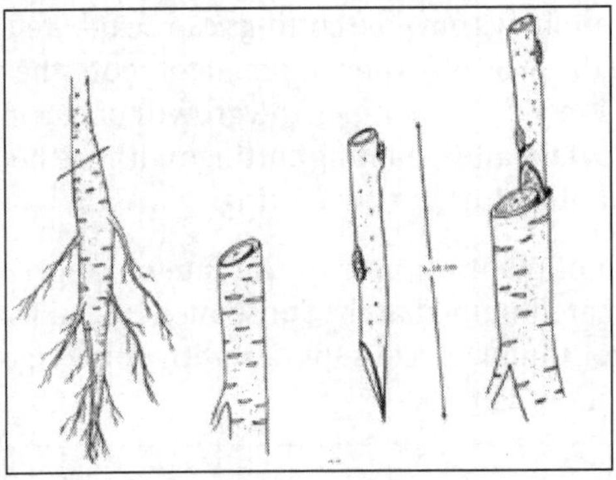

Root Grafting.

☆ **Bud Grafting:** Bud grafting involves removal of only one bud and bark and grafting it on the other plant. Following types of bud grafting can be used.

- ❖ **Patch Budding:** A portion of the bark with an active bud is removed from the stem as scion and correspondingly a portion of the bark along with bud is removed from another stem (stock) to which bud grafting is to be made. To this required portion of stock, a bud of the scion is inserted and properly tied using polythene tape.
- ❖ **T-Budding:** A "T" shaped insertion is made on the barks of the stock at the nodal region and a slit is formed by opening the cut ends. A "T" shaped bud from the selected plant as scion is inserted in this (stock) slit and properly bandaged. This type of grafting is also called as shield budding
- ❖ **Shoot Grafting:** It is followed to renovate the old plants.
- ❖ **Wedge Grafting:** This is similar to root grafting except that in this case "V" cut is given on a shoot (stock). Prior to this, the plant is pruned at a required height and the scion is inserted later.

❖ **Crown Grafting:** This is similar to the above, except that more than one scion is inserted into the stock.

5.2.2.3 Layering

If the layering is done with the shoots, it is known as air layering. In this case, roots are allowed to develop at the nodes of the stem while it is attached to the parent plant. There are two types of layering *viz.* air layering and trench or ground layering.

☆ **Air Layering:** In this method, the bark and some portion of wood is removed from all around a shoot. Besides this, mosses (a kind of a small plant) are wrapped around the shoot at the point of cut. Subsequently, a polythene sheet is wrapped around the shoot. The two cut ends of the polythene are closed with a tag. It is to be kept in mind that shoots on which layering have been done should not be disturbed for 4-6 weeks. The rooted shoots are then removed from the parent plant and planted in nursery beds. Figure gives a view of air layering and after seeing the figure it may be easy to do layering.

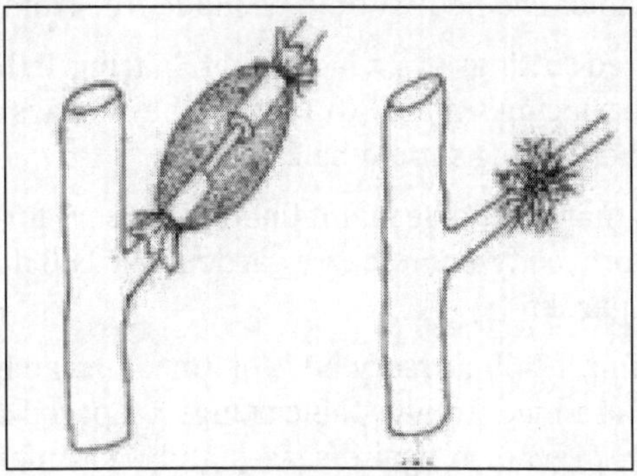

Air Layering.

☆ **Ground or Trench Layering:** In this method, the branches of a plant are placed horizontally in the base of the trench and covered along with its entire length filled with manure and soil. New shoots which develop from the buds along with the stem grow upwards through the soil. The shoots along with the roots are then planted as new plants. This method of grafting popular in Japan.

Trench or Ground Layering.

5.6.4 Preparation of Nursery

Nursery is a piece of well ploughed and manured land where mulberry cuttings are planted in a proper space to get well developed young rooted plants called saplings.

It is made with an intention of establishing mulberry plantation by using saplings to get a uniform plantation with less mortality. Since, direct plantation of mulberry with the cuttings has a high rate of mortality, especially in a place with lack of irrigation facility, the use of mulberry saplings is encouraged. Hence, raising a nursery before taking up a plantation has a very significant meaning.

To raise a nursery, it is highly essential to prepare the land with a lot of care. The land must be ploughed deep (> 30cm) with tractor drawn mould board plough two times depending upon the soil texture.

Farmyard manure should be applied at the rate of 10 MT/acre and mixed well with the soil by ploughing. In case of black soil, application of sand @ 10-20 MT/ acre is essential. Raised beds of convenient size, preferably of 300cm x 120cm x 10 cm breadth, length, and height, respectively, with irrigation channel all-round the bed is more suitable. The beds should be made free from weeds by digging.

Planting the cutted cuttings by giving proper Spacing. If the sapling developing for tree type then the spacing should 30x10 cm if the cuttings are planting for bush type purpose then the spacing should be 20x10 cm.

Number of days that should be maintained at nursery for transplanting is 240 days from planting for tree type mulberry garden and 180 days from planting for bush type mulberry garden

In one acre of land, 1,065 nursery beds of 3m x 1m can be prepared. As soon as the nursery beds are made ready, the cuttings prepared should be planted in the nursery beds. To prevent fungal diseases and pest infestation, 0.1 per cent Bavistin and 0.2 per cent Dimethoate (Rogar) spraying after 30-40 days of planting is suggested. Appropriate measures must be taken to control any incidence of diseases and pests.

After 4 months of planting, the saplings will attain about 90 to 100 cm height and will be ready for transplantation. Saplings should be uprooted carefully without damaging the roots. For easy uprooting of saplings, the beds should be irrigated two days before uprooting. The saplings should be planted as quickly as possible to get high rate of survival.

If the saplings are not planted in the main field immediately after uprooting due to some unavoidable reasons, then saplings should be preserved by covering with wet gunny cloth. Transportation of saplings for longer distance should be avoided or preferably done during cooler hours.

Before taking up the plantation, the saplings should be checked for root-borne diseases and root rot and the infected saplings, if any, should be discarded by burning. By following standard nursery technique as described above, it is possible to plant about 1.60 lakh cuttings in one acre and about 1.28 lakh good saplings @ 80 per cent survival can be obtained.

Chapter 6

Advanced Micropropagation Methods

6.1. Introduction

The main focus of mulberry breeding is to improve leaf productivity as it alone contributes more than 38.2 per cent to the sericulture productivity. However, it is not easy to improve the leaf productivity as it is a multifactorial trait determined by a number of associated characters such as plant height, number of branches, leaf retention capacity, nodal length, leaf size and weight, total biomass *etc.*

High heterozygosity and inbreeding depression hinder the development of inbreds, hence, directional breeding failed to make much progress. Therefore, the heterozygous parents are used to generate F1 progenies, which are then subjected to different evaluation and selection procedures to identify the best one. This type of breeding system bars the possibility of introgressing genes of desirable traits from wild relatives/species due to genetic drag and subsequent difficulty in eliminating the undesirable traits that come along with it.

Under such circumstances, the feasible means of improving specific traits without disturbing the current trait combinations is adoption of biotechnological tools like transgenesis, which enable introduction/over expression of desirable genes, or knocking out undesirable genes (RNA interference).

Mulberry, being a tree with high heterozygozity, poses difficulties on improving traits of economic importance through conventional breeding and selection. Environmentally less influenced and developmentally stable molecular markers provide reliable tools for the breeders to characterize the germplasm and to select parents and offsprings through marker assisted selection. Thus, it would be prudent to use biotechnological tools to harness the vast benefit mulberry offers to mankind.

6.2 Tissue Culture in Mulberry

Tissue culture technique in mulberry has developed and ramified into different areas such as micropropagation, callus culture, organogenesis, screening of genotypes for stress tolerance, induction of polyploids, cryopreservation, transgenesis and others. Following are some of the significant contributions tissue culture made in mulberry.

6.2.1 Micropropagation

Mulberry can be vegetatively propagated through stem cuttings, grafting or budding. However, success of these methods depends on a number of factors such as genetic makeup of the plant, age and physiological conditions of the parental cutting, climatic conditions and others. Additionally, newly developed mulberry varieties cannot immediately be propagated through stem cuttings as at least 6-7 month maturity is required to make the cuttings from the parental plant.

Micropropagation on the other hand, allows multiplication of the plant in a short period under the controlled conditions. Further, in conventional method of propagation through stem cuttings, each stem cutting produces only one plant, whereas in micropropagation thousands of plants can be produced from a single plant piece Moreover micropropagation can provide plantlets throughout the year irrespective of seasonal variations,; It is, thus, an efficient and cost effective tool for rapid multiplication of mulberry in a relatively shorter time and space.

Micropropagation also facilitates production of virus-free plants from the apical meristematic tissues. However, success of micropropagation is dependent on a number of factors among them genetic makeup, age and origin, physiological and pathological conditions of the explants, media composition and culture conditions are considered key factors.

Shoot tips and dormant axillary buds were found suitable for mulberry micropropagation. A number of factors significantly effect microprogation in mulberry, among them composition of media is an important factor. Among the different media compositions and hormones tested, MS medium containing 2 mg L^{-1} 6-benzylamino purine (BAP) is the best for shoot multiplication. However, it is also important to note that 6- benzyladenine (BA) above 2 mg L^{-1} is inhibitory for shoot initiation and multiplication. Among different pH levels, 5.6-5.8 is optimum for shoot multiplication. Agar in the culture medium also plays important role in the success of micropropagation. Among various concentrations, 0.8 per cent agar is the best and above 1.0 per cent found reducing shoot growth.

Table 1: *In vitro* Plant Regeneration in Mulberry (*Morus* spp.) during the Last One Decade

Sl.No.	*Species*	*Tissue*	*Mode of Regeneration*
1	M. alba	Shoot tips, Nodal segments	Micropropagation
2	M alba	Nodal segment, Shoot tips	Micropropagation
3	M. alba	Cotyledons	Micropropagation
4	M. alaba	Leaves	Micropropagation
5	M.alba	Nodal segments	Micropropagation
6	Morus spp	Side bud	Organogenesis
7	M. alba	Nodal segments	Oraganogenesis
8	Morus spp.	Nodal explants	Oraganogenesis
9	M. nigra	Nodal segments	Micropropagation
10	M. alba	Nodal segments	Micropropagation
11	M. alba	Nodal segments	Micropropagation
12	M. macroura	Nodal segments	Micropropagation
13	M. indica	Nodal segments	Micropropagation

6.2.2 Organogenesis in Mulberry

Organogenesis is a complex phenomenon involving de novo formation of organs. Successful organogenesis depends on a number of factors which include appropriate selection of explants, age of the explants, media compositions, specific growth regulators, genotype, sources of carbohydrate, gelling agent, and other physical factors including light, temperature, humidity and other factors. Depending on these factors plant regeneration may occur either directly or indirectly. In direct organogenesis, plants develop directly from the explants without formation intermediate callus while in indirect organogenesis plant develops via callus formation.

Again, callus induction depends on a number of factors such as nature of explants, genotype, medium and its composition. A variety of explants has been tested to initiate callus formation in mulberry. Indirect organogenesis often creates variations among the regenerated plants, a phenomenon called somaclonal variation, which often facilitates isolation of useful plants. Direct organogenesis from explants especially from cotyledons and leaf segments has great potential for transgenesis. In mulberry, direct plant regeneration from leaf explants was first reported by Kim *et al.* (1985) and later by Yamanouchi *et al.* (1999).

6.2.3 Somatic Embryogenesis

Somatic embryogenesis provides a valuable tool to enhance the pace of genetic improvement of commercial crop species (Stasolla and Yeung, 2003). Several investigating groups attempted induction of somatic embryos in mulberry but the rate of success is less. Due to the difficulty in hormonally controlling the formation of adventitious shoots and roots in mulberry, somatic embryogenesis has not been developed as it is in many other crop plants. Thus, concerted efforts are needed to make somatic embryogenesis successful in mulberry.

Figure 1: Micropropagation in Mulberry.

A: Nodal explant, B: Shoot from the explant; C: Multiple shoots growing from explant; D: Rooting of the micropropagated plant; E: Hardening of the micropropagated plant.

Table 2: Direct Plant Regeneration from Mulberry Leaf, Cotyledon and Hypocotyls Explants

Explant	*Medium +PGR*	*Species*
Leaf	MS+ BA, MS+ BA+ NAA	Morus sp.
	MS + BA + BR	M. alba
	MS + TDZ + TIBA	M. alba
	MS + BA	M. indica
	MS+TDZ + IAA + $AgNO_3$	M. indica
	Mod. MS + TDZ	M. alba; M. latifolia
	ModMS+IAA+TDZ+$AgNO_3$	M. indica
	MS + TDZ + IAA + $AgNO_3$	M. indica
Cotyledon/hypocotyls region	MS + IAA + BAP	M. indica
	MS+TDZ	M. alba
	MS+TDZ+IAA+$AgNO_3$	M. indica
	ModMS+IAA+TDZ+$AgNO_3$	M. indica

$AgNO_3$- Silver nitrate; BA-Benzyl adinine; BR- Brassinosteroid; IAA-Indole acetic acid; MS-Murashige and Skoog (1962) medium; NAA-α-Napthalene acetic acid; TDZ- Thidiazuron.

6.2.4 Haploid Production

Haploid plants being gametophytic in origin possess only half the normal number of chromosomes as present in the parent. They can be used to produce homozygous lines, which are invaluable for any breeding programmes especially for tree crops with longer generation cycle and high heterozygosity. Since the first successful report on regeneration of haploid plants from pollen grains of the cultured anthers of datura, this technique has been extensively used in most of the agriculturally important plant species. However, only limited success could be obtained on tree species. In Mulberry, though anther culture was first attempted by Shoukang *et al.* (1987) and till date no plants could be regenerated. However, no further report on haploidy is available in mulberry, though doubled haploidy is of much use in mulberry breeding.

6.2.5 Protoplast Isolation, Culture and Regeneration of Plantlets

Somatic hybridization through protoplast fusion has opened a new avenue for developing new characteristics, which are not possible through conventional breeding. There are only a few reports dealing with plant regeneration from protoplasts in mulberry. A combination of 2 per cent cellulase, 1 per cent macerozyme and 0.5 per cent macerase is found optimal for better isolation of viable protoplast. Protoplast fusion in mulberry was successfully achieved using chemical fusogen and electro-fusion. Although protoplast isolation and regeneration were achieved, development of somatic hybrids in mulberry could not be achieved. Hence, efforts in this end need to be continued.

6.2.6 Other Applications of Tissue Culture in Mulberry Screening for Stress Tolerance

Since salt tolerance in plants is a complex phenomenon involving morphological, physiological and biochemical processes, screening of genotypes for salt tolerance need to be done in such conditions where the influence of external factors is minimal. Maintenance of uniformity of salinity across the field and seasons is difficult, therefore, screening of the plants under in vitro is considered as an ideal option, where most of the environmental conditions can be controlled. Vijayan *et al.* (2003) using axillary buds of 63 mulberry germplasm accessions maintained at the Central Sericultural Research and Training Institute, Berhampore, West Bengal, India isolated salt tolerant genotypes by surface sterilizing the nodal explants and culturing on MS medium (Murashige and Skoog, 1962) supplemented with 2 mg L^{-1} BAP, 30 mg L^{-1-1} sucrose and 0.0 per cent to 1.0 per cent NaCl (Figure). Genotypes showing early sprouting and better growth rate in 1.0 per cent NaCl were selected as salt tolerant. Out of the 63 accessions tested, 16 sprouted in 1.0 per cent NaCl, and 13 of them survived till 30 days and these genotypes were considered as salt tolerant. For rooting assessment, in vitro grown shoot apices (1-2 cm in length) were cultured on MS medium containing 2.6 μM NAA, 30 g l^{-1} sucrose and 0.1, 0.2, 0.3 and 0.4 per cent NaCl. Out of 63 genotypes tested, only five genotypes could develop roots at 0.2 per cent NaCl. Five genotypes, which showed better root and shoot growth in both experiments, were further tested in pot culture under different salt levels to confirm the efficacy of the in vitro screening. The study while confirming the higher salt tolerance of the selected genotypes also demonstrated the efficacy of in vitro screening to isolate salt tolerant genotypes in mulberry. This method is more economical, efficient and less time consuming for screening large number of mulberry accessions for salt tolerance.

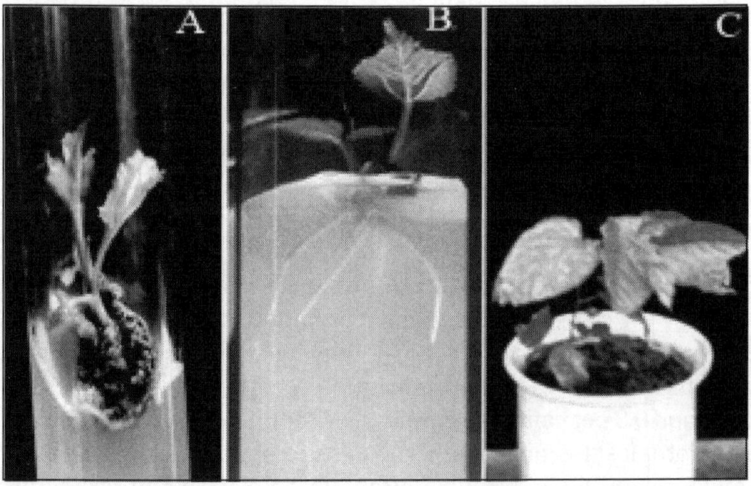

Figure 2: Direct Plantlets Regeneration from Leaf Explants.

A: Shoot formation from leaf; B: Root formation from the shoot; C: Whole plant regenerated from leaf.

6.2.7 Induction of Tetraploidy

In general, the mulberry is propagated through vegetative means. Hence, sterile high yielding varieties/cultivars do not pose any problems for their true to type multiplication. Triploidy in mulberry is considered as the optimum level of ploidy because triploids show several advantages over plants of other ploids. Triploids are superior in leaf yield, stress resistance and chemical components of the leaf (Yang and Yang, 1989). Considering these advantages, tetraploids are developed from diploids by colchicine treatment of the growing shoots. In this method, small cotton pads soaked with 1.0-2.0 per cent colchicine solution is applied over the growing buds for 2-3 consecutive days. Though this method is easier to apply, it suffers from quick drying of the cotton pad, excessive loss of colchicines and difficulty in maintaining the uniform concentration of the colchicines solution. Application of colchicine in vitro solves most of these problems and also makes the system more economic. Chakraborti *et al.* (1998) cultured apical buds of field grown mulberry on MS medium supplemented with BAP (2 mg L^{-1}) and four concentrations of colchicine (0.0, 0.05, 0.1 and 0.2 per cent). It was observed that the optimal level of colchicine was 0.1 per cent as it produced 39.4 ± 4.8 per cent tetraploids. Higher concentrations of colchicine reduced the survivability of the buds and low concentration reduced the percentage of tetraploid formation. It is further observed that in vitro application of colchicine was 80.8 per cent more efficient than the cotton pad method. Additionally, in vitro application of colchicine is more cost effective as the same medium can be used for at least 4 repeated treatments without reducing the efficiency of the colchicine to induce tetraploidy.

Another method of getting triploidy in mulberry is to culture the endosperm because in angiosperm, endosperm is a triploid tissue formed via double fertilization. In mulberry, Thomas *et al.* (2000), for the first time, successfully developed triploids from endosperm of the variety S36. Endosperm tissue isolated from young fruits was cultured on MS medium supplemented with 2,4-D (5 μM) and different concentration of BA, Kn, TDZ, IBA, NAA, gibberellic acid (GA3), along with tomato juice (TJ), yeast extract (YE), casein hydrolysate (CH) and coconut milk (CM). Shoot buds developed when the callus was subcultured on a medium containing a cytokinin or a combination of cytokinin and NAA. The best result for callus induction (70-72 per cent) was obtained on MS medium with BAP (5 mM), NAA (1 μM) and CM (15 per cent) or YE (1000 mg/l). The maximum number of shoot formation was on the medium containing TDZ (1 μM), or BA (5 μM) and NAA (1 μM). On cytological investigation, it was found that all the ten plants examined were triploids with 42 chromosomes. This clearly indicated the efficiency of the protocol and it could be used for developing more triploids in mulberry.

6.3 Synthetic Seeds

Synthetic seeds are the encapsulated somatic embryos, which functionally mimic zygotic seeds and can develop into seedlings under sterile conditions. In a broader sense, it would also refer to encapsulated buds or any other form of meristems, which can develop into plants.

In mulberry, synthetic seeds are produced mostly by encapsulating the apical/ axillary buds or somatic embryos with 3-5 per cent sodium alginate and 100mM calcium chloride solution as complexing agent. Sodium alginate solution is mixed with culture medium containing all necessary ingredients essential for proper growth. Patnaik *et al.* (1995) successfully developed this technology for artificial seed sysnthesis in mulberry. However, adoption of this technology for mulberry propagation was limited to a few species of M. indica.

Researchers have explored the possibility of using in vitro derived vegetative propagules for synthetic seed production since it was found difficult to develop somatic embryos in mulberry. Shoot cultures established from axillary buds on Linsmaier and Skoog's basal medium (LSBM) supplemented with BAP (4 mg L^{-1}) and TIBA (1 mg L^{-1}) used for encapsulation. Sodium alginate and carboxy methylcellulose were added in the range of 2, 4, 6 and 8 per cent (w/v) to liquid initiation medium separately. For complexation, 25 x 103 μM, 5 x 104 μM, 75 x 103 μM and 10 x 104 μM calcium chloride solutions were prepared using distilled water. Gel complexation was done by mixing the axillary buds with hydrogels, dropping these into different concentrations of calcium chloride solution and incubating in orbital shaker for different time intervals (20, 30, 40, 50 and 60 min) to obtain uniform beads. The sodium alginate and carboxy methylcellulose embedded axillary buds were collected using a sterilized tea strainer and rinsed 2-3 times in sterile water to remove traces of calcium chloride.

The synthetic seeds were tested for their conversion potential under in vitro and ex vitro conditions. Although different concentrations of sodium alginate were used, the best results were obtained at 4 per cent alginate. Although somatic embryos could not be developed easily in mulberry, dormant axillary buds proved to be an ideal material for the synthesis of synthetic seed, it can be used for cryopreservation of germplasm accessions.

6.4 Cryopreservation of Germplasm

The high heterozygosity hinders conservation of mulberry germplasm through seeds as the progenies from such seeds are heterogenous in nature and getting true to the parental type is difficult. Thus, mulberry germplasm is conserved as ex situ germplasms, which is laborious, needs huge resources, and is in a risk of destruction by natural calamities, pests and diseases. Thus, safe alternative methods with economically viability need to be explored. Cryopreservation is one such alternative wherein plant materials are stored at ultra-low temperatures (-196°C) in liquid nitrogen. At this temperature all the metabolic activities of the cell including divisions remain arrested; hence, the material remains unaltered for long period.

Two different cryopreservation techniques in vogue are the classical one and the modern vitrification (Engelmann, 2000). In classical cryopreservation technique, the plant material is cool down slowly at a controlled rate of 0.1-4°C/min to about -40°C and rapidly immersed in liquid nitrogen. In vitrification, plant material is physically or osmotically dehydrated and is subsequently subjected to ultra-rapid freezing resulting in vitrification of intracellular solutes, *i.e.* formation of an amorphous glassy structure without occurrence of ice crystals. Although different plant materials are used for cryopreservation, the most appropriate material for cryopreservation of mulberry is winter buds (Niino, 1995).

Genetic constitution of the plant material also influences significantly the success and duration of the preservation (Rao *et al.,* 2007). Cryopreservation techniques have been standardized for M. indica, M. alba, M. latfolia, M. cathayana, M. laevigata, M. nigra, M. australis, M. bombycis, M. sinensis, M multicaulis and M. rotundiloba species.

The general procedure for cryopreservation of shoot tips is that the shoot segments are pre-frozen at -3°C for 10 days, -5°C for three days, -10°C for one day and -20°C for one day and immersed into liquid nitrogen. The cryopreserved material is brought back to the ambient temperature via thawing. During this process, the material is warmed either slowly at 0oC in air for more than 3 h or rapidly in agitated water at 37oC for 5 min. After thawing, surface-sterilize the bud with 70 per cent ethanol for 1 min and sodium hypochlorite solution (0.5 per cent effective chlorine concentration) for 20-30 minute and rinse it thoroughly with sterile distilled water for three to four times before culturing on MS medium containing all necessary supplements. Niino *et al.* (2006) obtained 65 per cent survival of the buds cryopreserved at -135oC for 5 years. The Surviving buds resume growth within 5 days and develop shoots without intermediary callus formation. It is also found that partial dehydration of the bud up to 38.5 per cent before pre-freezing at -20 °C improves the recovery rate. Encapsulation of winter-hardened shoot tips with calcium alginate coating was also tested successfully. It is concluded from different experiments that dormant buds of mulberry can be cryopreserved for 11 years without reducing the viability of the buds.

6.5 Major Problems Associated in Mulberry Tissue Culture

Problems in bud culture High rate of contamination is one of the major problems that limit the use of mature field grown stem cutting for in vitro studies, especially for flowering and in vitro fertilization. Since the field grown plants are often heavily loaded with various epiphytic and endophytic microbes, it is very difficult to remove them, especially from the lenticels. Further, it is also noted that occurrence of microbial contamination depends to a great extent on genotype, maturity of the explants and seasons. During rainy season, heavy bacterial infection was observed. Patnaik and Chand (1997) reported that an initial thorough washing of field grown nodal explants with running water for 1-2 h followed by treating it with 5 per cent (v/v) liquid detergent and 7 per cent sodium hyphochlorite for 8-10 minutes are essential before doing the normal surface sterilization with 0.1 per cent aqueous (w/v) mercuric chloride to contain contaminations. Vijayan *et al.* (2011b) found that when mature nodal plants are used for micropropagation, fungal growth takes place after 20-30 days from the lenticels and scale leaves. In order to avoid this contamination, the young shoots should be subcultured as soon as they emerge from the buds of the explants collected from the field.

6.5.1 Problems in Callus Culture

Oraganogenesis in mulberry is also not devoid of problems as often excessive blackening of the callus occurs due to exudation and oxidation of the phenol contents. The oxidative products of phenolic compounds lower the pH of the culture medium, thereby affecting the callus growth. This problem can be limited to an extent by the addition of activated charcoal and or silver nitrate in the culture medium. Regular subculturing at an interval of 20-25 days is another way of controlling this problem. Another major problem is the formation of roots from the callus before the shoot formation. Once the root formation starts shoot formation becomes difficult from the same callus. This serious problem should be controlled by hormonal manipulation such that the effect of cytokinin should not be less than that of auxin in the initial stages of organogenesis.

6.5.2 Problems Associated with Acclimatization and Field Transfer

During in vitro culture, plantlets grow under control conditions such as humidity, irradiance, temperature and other factors. The plantlets cultured in vitro in a water-saturated atmosphere generally wilt rapidly when transferred to normal greenhouse or field conditions and also are highly prone to microbial attacks, if adequate care is not taken. One of the major reasons for wilting of the transferred plant is the rapid loss of water from the leaves due to poorly developed stomata and epicuticular wax. Poor contact of roots with the soil also limits water up take. Although it is quite common that leaves with normal stomatal structure replaces the older ones, it that plantlets die during the acclimatization process. To minimise such casualties, it is essential to provide with a regulated temperature, humidity, irradiance, CO2 concentration and air flow rate during acclimatization. Initially the in vitro developed plants can be kept inside the culture room under sterile conditions. Then, gradually it can be shifted to green house where humidity and temperature are under control. After a couple of months, the grown plants can be shifted to shady places in the field, thereafter it can be transplanted to field.

6.6 Genetic Engineering in Mulberry

Although conventional plant breeding has contributed significantly by developing several high yielding mulberry varieties (Table 3), high heterozygozity and long generation cycles and inbreeding depression prevent introgression of traits from wild species and relatives. Transgenesis is one such technique that enables direct transfer of genes of interest. Out of the two popular gene transfer techniques *viz.*, particle bombardment and Agrobacterium tumefaciens-mediated transformation, the latter received much attention in mulberry due to the easiness and efficacy.

Bhatnagar *et al.* (2003) generated transgenic mulberry expressing GUS genes. In the process, it is found that A. tumefaciens strain LBA4404 was more infective than the other strains. Nearly 90- 100 per cent transformation was obtained with LBA4404, 70-75 per cent with GV2260 and 25-35 per cent with A281. Among the plasmids, pBI121 and pBI101:Act1 were most efficient (100 per cent). The transgenic mulberry over-expressing HVA1 from barley showed comparatively better growth under abiotic stresses. The transgenic plants with Osmotin gene under the constitutive expression of the CaMV 35S promoter and a stress-inducible promoter rd29A also expressed high salt, drought and cold stress tolerance. Thus, transgenesis in mulberry seems to have come of the age.

In view of the above developments, attention should be directed to some of the burning issues of mulberry cultivation. Major diseases like powdery mildew (Phyllactinia corylea) and bacterial blight (Xanthomona campestris cv. Mori) are causing a crop loss of 20-30 per cent in the field. Therefore, sincere and serious efforts should be made to develop disease resistant transgenic plants against these diseases.

Table 3: High Yielding Mulberry Varieties Developed for Commercial Cultivation in India

Variety	*Regions/Zones*	*Cultural Condition*
V1	Karnataka, Andhra Pradesh, Tamil Nadu	Irrigated
Anantha	Karnataka, Andhra Pradesh, Tamil Nadu	Irrigated
Vishala	All sericulture regions of India	Irrigated
S1635	Karnataka, Andhra Pradesh, Tamil Nadu, Maharashtra, West Bengal, Uttar Pradesh, Bihar, Assam and other states in Central India	Irrigated
DD	Karnataka, Andhra Pradesh, Tamil Nadu, Maharashtra and Kerala	Irrigated
S36	Karnataka, Andhra Pradesh, Tamil Nadu and Kerala	Irrigated
S1	West Bengal, Uttar Pradesh, Bihar, Assam and Central India	Irrigated, Rainfed
S799	Sericultural regions in central India	Irrigated
S13	Red soil of Karnataka, Andhra Pradesh, Tamil Nadu	Rain-fed
S34	Black soils of Karnataka, Andhra Pradesh, Tamil Nadu	Rain-fed
BC259	Hilly Eastern areas of India	Rain-fed
S146	Central India, Doon valley, Himachal Pradesh and Hilly Eastern areas of India	Rain-fed
Tr-10	Central India, West North Jammu, Doon valley, Himachal Pradesh and Hilly Eastern areas of India	Rain-fed
Chak majra	Western North, Jammu and Himachal Pradesh	Rain-fed
Chinese white	Western North, Jammu and Western Himachal Pradesh	Rain-fed

6.6.1 Molecular Markers

From the research during the last couple of decades clearly demonstrates that four major types of molecular markers have been used in mulberry for various purposes which includes elucidation of genetic diversity among germplasm accessions, relationship among populations, cultivars, land races and species.

6.6.2 Gene Transfer Technique in Mulberry

6.6.2.1 Vector-Mediated Gene Transfer

Agrobacterium-Mediated Gene Transfer

Different vectors used in gene transmission in mulberry, among them the Ti plasmid of Agrobacterium tumefaction's has been widely used because it has natural ability to transfer T-DNA of its plasmids into plant genome upon infection of cells at the wound site. Ti plasmids are used as gene vectors for delivering useful foreign genes into target plant cells and tissues. The foreign gene is cloned in the T- DNA region of Ti plasmid. To transfer the foreign DNA in host plants, leaf discs, embryogenic callus or other suitable explants are collected and infected with Agrobacterium carrying recombinant disarmed Ti plasmid vector. (Reviewed by Shabir *et al.,* 2013). Agrobacterium-mediated transformation of HVA1 demonstrated the gain of tolerance against drought, salt and cold stress; in this study they proved that Arabidopsis rd29A promoter can function in mulberry (Checker *et al.,* 2012).

Achievements

Delhi University group has developed transgenic mulberry expressing HVA1, LEA3 group stress responsive gene from barly.

Over expression of HVA1 gene from barley generates tolerance to salinity and water stress in transgenic mulberry. The HVA1 gene from barley encodes a group 3LEA protein and is induced by abcissic acid (ABA) and water deficit conditions. Over expression of HVA1 in mulberry under a constitutive promoter via Agrobacterium mediated transformation. Molecular analysis of the transgenic plants revealed the stable integration and expression of the transgenic in the transformants. The transgenic plants showed better cellular membrane suitably, photosynthetic yield, less photo-oxidative damage and better water use efficiency as compared to non-transgenic plants under both salinity and drought stress. Amongst the lines analyzed for stress tolerance transgenic lines ST8 was relatively more salt tolerant, ST30 and ST31 was more drought tolerant (Lal *et al.,* 2008).

Expression of Arabidopsis SHN1 in Indian Mulberry (*Morus indica* L.) Increases Leaf Surface Wax Content and Reduces Postharvest Water Loss. Leaf surface properties regulate direct water loss through the cuticular layer. Leaf surface waxes, contribute for cuticular resistance and protect mesophyll cells from desiccation (Sajeevan *et al.,* 2017).

DNA Viruses as Vectors

The genomes of Caulimoviruses such as cauliflower mosaic virus (CaMV) and geminivirus such as tomato golden mosaic virus (TGMV) are double- stranded DNA which makes these viruses as potential transformation vectors. Methotrexate resistant dihydrofolate reductive (DHFR) gene of *E. coli* has been successfully cloned into an intergenic region of CaMV. This engineered CaMV was used to infect turnip plants (Reviewed by Shabir *et al.,* 2013).

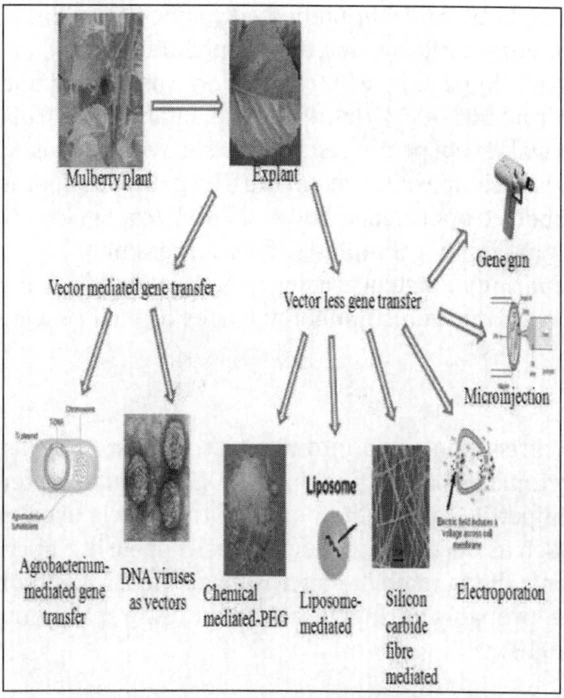

Gene Transfer Methods.

6.6.2.2 Vector-Mediated Gene Transfer

Chemical Mediated Gene Transfer

Cells or protoplasts can be stimulated to take up foreign DNA using some chemicals. Polyethylene glycol (PEG) is the most commonly used chemical for this purpose. It helps in precipitation of DNA which can then be taken up by the calls through the process of endocytosis (Roland bilang, *et al.*, 1994).

Liposome-Mediated Gene Transfer

Plasmid containing foreign desired gene can be enclosed in small lipid bags called liposomes, which can then be fused with protoplasts using chemicals like PEG.

Silicon Carbide Method

In this method, fibres of organic material like silicon carbide are used for gene transfer. These fibres, when mixed with plasmid DNA and plant tissue or cells, help in penetration of the foreign DNA into the plant tissue.

Electroporation

Gene transfer by electroporation into protoplasts isolated from mulberry for callus induction, explants were excised from the roots of mulberry seedlings grown in a medium containing benzyl aminopurine, dichlorophenoxy acetic acid and thidiazuron. The isolated calli were repeatedly subcultured in the liquid medium. Protoplasts were enzymatically isolated from clumped cells and transfer of the β- glucuronidase (GUS) gene by electroporation was carried out at various pulse voltages ranging from 500-750 V/cm with a capacitance of 330 μFd. A typical result showed that survival rate of protoplasts decreased with increasing pulse voltage in electroporation. Transient expression of the GUS gene was operative in 20 - 30 per cent of protoplast electroporated at 500 and 750 V/cm. which is the first time this has been done in mulberry protoplasts. This success may open new possibilities in the analysis of various constructs using genes isolated from mulberry trees and in the production of transgenic mulberry clones as well (Reviewed by Shabir *et al.,* 2013).

Microinjection

The DNA is directly injected into plant protoplasts or cells (especially into the nucleus or cytoplasm) using fine tipped (0.5-1.0 micrometer diameter) glass needle or micropipette. This method of gene transfer is used to introduce DNA into large cells such as oocytes, eggs and the cells of early embryo. The process is observed and controlled under the microscope. The DNA is then integrated into the plant genome-probably during the cells own DNA repair processes (Reviewed by Shabir *et al.,*2013).

Particle Bombardment Gun

As the term denotes, it shoots foreign DNA into plant cells or tissue at a very high speed. This technique is also known as particle bombardment, the apparatus consists of a chamber connected to an outlet to create vacuum. At the top, a cylinder is temporarily sealed off from the rest of chamber with a plastic rupture disk. Helium gas flows into the cylinder. A plastic microcarrier is placed close to rupture disk. It contains DNA coated tungsten particle, the microscopic pellets (*i.e.* coated microprojectiles). When to work the apparatus is placed in Laminar flow just to maintain sterile conditions. The target cells/tissue is placed in the apparatus. A stopping screen is put between the target cells and microcarrier assembly. Helium gas is flown in the cylinder at high velocity. When pressure of cylinder exceeds the bursting point of plastic disk, it gets ruptured. Helium shock waves propel the plastic microcarrier containing DNA coated micro pellets. The stopping screen allows the micro pellets to pass through and deliver DNA into target cells. The transformed cells are regenerated onto nutrient medium. Major improvements in gene delivery into mulberry leaf cells by particle inflow gun. The efficiency of gene delivery into mulberry leaf cells using a particle inflow gun for the production of β-glucuronidase (Yukie *et al.,* 1999).

Transgenic Mulberry

Step 1: Nucleic acid (DNA/RNA) Extraction: The initial step is the disruption of the desired organism, which may be viral, bacterial or plant cells, in order to extract the nucleic acid. After a series of chemical and biochemical steps, the extracted nucleic acid can be precipitated to form a thread-like pellet referring to the DNA/RNA.

Step 2: Gene Cloning: Upon DNA extraction, all DNA from the desired organism is extracted at once. Through gene cloning, the desired gene/s can be isolated from the rest of the DNA extracted, which is then mass-produced in a host cell to make thousands of copies of the desired gene.

Step 3: Gene Design and Packaging: Once the gene of interest has been cloned, it has to be linked to pieces of DNA that will control how the gene of interest will work once it is inside the plant genome. These pieces of DNA will switch on (promoter) and of the expression of the gene inserted. Gene designing/packaging is done by replacing an existing promoter with a new one and incorporating a selectable marker gene.

Step 4: Transformation: Using the gene gun or Agrobacterium-mediated transformation. The main goal in any transformation procedure is to introduce the gene of interest into the nucleus of the cell without affecting the cell's ability to survive. If the introduced gene is functional, and the gene product is synthesized, then the plant is said to be transformed. Once the gene inserted is stable, inherited and expressed in subsequent generations, then the plant is considered a transgenic.

Step 5: Backcross Breeding: This is done by crossing the transgenic plant with elite lines using conventional plant breeding methods. This enables the combination of the desired traits of the elite parents and the transgenic into a single line. The offspring are repeatedly crossed back to the elite line to obtain a high yielding transgenic line.

Mulberry transgenic plants expressing *AtSHN1* displayed significant improvement in leaf moisture retention capacity even 5 h after harvest and there was a slow degradation of total buffer soluble protein in detached leaves compared to wild-type. Silkworm bioassay did not indicate any undesirable effects on larval growth and cocoon yield. This clears that expression of *AtSHN1*, can increase surface wax load and reduce the postharvest water loss in mulberry.

Chapter 7

Methods of Irrigation, Manuring and Pruning

7.1 Irrigation: An Introduction

Of all the major inputs in mulberry cultivation, irrigation plays a vital role in increasing the leaf yield and quality as it helps in better utilization of the nutrients absorbed by the plants. Irrigation water should be used carefully without wasting this precious input.

Mulberry plants with good irrigation show luxuriant growth with succulent leaf, which is relished by the silkworm. If there is scarcity of water, the growth of the mulberry plants is restricted and they become poor in yield and quality. Normally, about 120-150 mm well-distributed rainfall in a month is required for the luxuriant growth of mulberry. But, such conditions rarely exist. Hence, it is very essential to supplement the required quantity of water by irrigation from time to time to maintain optimal soil moisture.

The frequency of irrigation for the mulberry crop depends upon several factors such as soil type, agro-climatic conditions and growth phase of the plant. Normally, the frequency of irrigation varies from 10-12 days interval in clayey loam soil and 7-10 days interval for sandy loam soils. For each irrigation, one and half inch of irrigation water (approximately 33,000 gallons) per acre is required.

7.1.1 Factors Affecting Frequency of Irrigation

☆ **Humidity:** In rainy season, the humidity is high and rains may be received just when the crop is in need of water. In such case, some irrigation turns could be stopped and frequency may be extended to 20 days. During winter season, also the frequency will be longer than in summer because of less evapotranspiration, dewfall, nighttime humidity, and less sunshine. The frequency may therefore be 15 to 20 days in winter and 6 to 8 days in summer. In summer irrigation, water is given more frequently and hence more frequency of irrigation in summer, medium in winter and less in rainy season.

- ☆ **Stage of Growth of Crops**: During certain stages particularly at flowering and fruit formation stages of crop requires much larges quantities of water than earlier stages. In earlier stage, even if a little less water than estimated daily use is provided, the crop will stand the strain without any harm, perhaps a slight moisture stress may encourage better root growth.
- ☆ **Type of Crop:** The frequency of irrigation will also depend up on the crop. A succulent leaf vegetable will require irrigation more often than cereal crop like Jowar. Crops which are doses of fertilizers need more water than those with a little or no fertilizers.
- ☆ **Soil Type:** Light soil requires more frequent irrigation than the loamy soils. Sandy loam soil need to be irrigated every fifth day while clay loam may be irrigated every tenth day. Time required to irrigate an area: The time required to irrigate an area depends up on magnitude of discharge, quantity of water applied, irrigation efficiency and area. The time required to irrigate an area is calculated by formula.

$$IQT = Ad$$

where,

I: Irrigation efficiency

Q: Discharge in cusec

T: Time in hours

A: Area in acres

d: moisture deficit in soil.

7.2 Types of Irrigation (Classification)

- ☆ Flood irrigation
- ☆ Surface irrigation
- ☆ Sub surface irrigation
- ☆ Sprinkler irrigation
- ☆ Drip irrigation

7.2.1 Flood irrigation

In flood irrigation, water floods over the soil surface and thus wets the soil.

Border Irrigation

With border irrigation, the water is diverted into a preconstructed bed and allowed to flow freely over the soil surface. The bed consists of an almost horizontal flow area and two earth borders to define the width of the bed.

Advantages

- Low energy costs, because the water flows by gravity over the field.
- The crop is not wetted, thus leaf and fruit diseases are reduced.
- Brackish soils can be leached with relative ease.
- Low capital input costs if the land is relatively level.

Disadvantages

- Considerable water losses may occur if the supply system is not properly designed and maintained.
- The system is very sensitive and small deviations from the design specifications can reduce application uniformity significantly.
- Not all crops can be grown in bordered beds and the type of crop to be planted may thus eliminate this irrigation method.
- The viability of such a system is mainly influenced by the extent of earthworks required for bed preparation.
- This type of system is relatively inflexible and difficult to alter once it has been installed.
- The system is relatively labour intensive.
- Very high management inputs are required.
- Unsuitable for soils with very high infiltration rates.

Semi-permanent: In this system the main and sub-main are fixed, usually buried under the ground where as the laterals are portables.

Solid straight system: These have enough laterals present to eliminate their movement from one place to another. The system is fixed at the beginning of system of season and remains through out the season for short and frequent irrigation.

7.2.2 Surface Irrigation

Water is applied directly to the soil from channel located at upper ridge of the field proper land preparation adequate control of water is necessary for uniform distribution of water border. The entire field is divided into strips separated by low ridge of the strip to lower in form of sheet guided by the low ridges. Border should have uniform gentle slope in direction of irrigation. Each strip is independently by turning stream of water at upper ridge. Suitability-suitable for close growing crops some row crop and orchards under favorable soil and topographic condition. Not recommended for extremely low or extremely high infiltration rate soils.

Advantage

- Easy construct and operate
- Person can irrigation more compares to check basin.
- If properly designed use uniform distribution and high water use efficiency.
- Large streams can be effectively used.
- If can provide excellent drainage (surface) if have proper outlet facility at the lower end.

Disadvantages

- Required precise land leveling
- Required large irrigation streams.

Furrow Method

Furrow is preferably used for row crops like maize, sugarcane, potato, groundnut and other vegetable crops. Water is applied in small furrows betureoil the row crops. Water infiltrated into soil and spread within the root zone. Large as well as small sized stream can be effectively used for irrigation. It also acids for safe disposal of excess water *i.e.* facilitates drainage. Only 1/5 to ½ of land surface is in contact with water (wet). There by reducing the evaporation losses. Method is specially situated to crops like maize which are sensitive to water in contact withtheir strength. The cost of land preparation is reduced and there is no wastage of land under field channels. In clay or deep clay soils shadow furrow are made along with guiding ridge to take care of soil cracking behavior such furrow are called corrugated furrow.

7.2.3 Subsurface Irrigation

Water is applied below the ground surface by maintaining artificial water table at some depth depends upon the soil characteristic and root zone of crop. Water moves through capillaries within soil to meet plant requirement deep trenches and underground piper are the two ways for subsurface irrigation.

Adaptability: Soils having low W.H.C. soil having very high-high infiltration rate. Soils surface method is not possible where sprinkle method of irrigation proves to be expensive.

Advantage: Evaporative losses are minimum.

Disadvantage: Salty water cannot be used.

7.3.4 Sprinkler Irrigation

It is methods in which water is spread into air and allowed to fall on the ground surface somewhat resembling. Water is forced under pressure through small nozzle/orifice which gets broken up to into droplets and fall back on the ground. Slow circular revolution is impacted to the nozzle uniformally covered the ground surface. The rate of application should not be more than the infiltration rate of the soil.

Adaptability of Sprinkler Irrigation

- ☆ Sprinkler irrigation can be adopted where land reveling is uneconomical and other method of surface irrigation can't carry out.
- ☆ Adapted to soils to pours highly avoidable or relatively impermeable which are difficult to irrigate by other methods like furrow, border *etc.*
- ☆ Where it is designed to go for frequent irrigation.
- ☆ It is designed to minimum cost towards labours, fertilizer, and irrigation.

Advantages of Sprinkler Irrigation

- ☆ It can be used for almost crops expect paddy and jute.
- ☆ System can be adopted under varied topographic condition and especially suitable to steep-slope and irregular topography
- ☆ Soils-method is particularly suited for sandy soils having high infiltration rate.
- ☆ It can eliminate surface run off of irrigation water (run off elimination)
- ☆ To protect the crop against frost and high temp and Savings in land construction of channel to the field.
- ☆ To reduce labour cost for irrigation as compared with surface method.
- ☆ Savings in land construction of channel to the field.
- ☆ It saves fertilizer and water as ferti-irrigation can be carried out.
- ☆ Land leveling is not essential for sprinkle irrigation.
- ☆ Gives higher water use efficiency

Limitations of Sprinkler Irrigation

- ☆ Not suitable for very fine texture soil (<4mm/her)
- ☆ Uneven distribution of water due to distortion by high water.
- ☆ More evaporation losses.
- ☆ Require clean, water free from debris sand slit and clay particles.
- ☆ Saline water cannot be used.
- ☆ Initial cost is high.

- High operating power is high (5-10kg/cm)
- Unsuitable climate condition sprinkling may be encouraging spread of disease.
- Ripening softy fruits need protection from the spray.

Systems of Sprinkler Irrigation

Based on Portability

- **Portable system:** In sprinkle irrigation system in which main line, sub main line, lateral and the pumping units all are the portable generally the pipers are made up of lightweight aluminum to facillate easy transportation such system can be shifted from place to another.
- **Semi-portable system:** This system is parallel to portable except that location of water source and pumping unit is fixed.
- **Semi-permanent:** In this system the main and sub-main are fixed, usually buried under the ground where as the laterals are portables.
- **Solid straight system:** These have enough laterals present to eliminate their movement from one place to another. The system is fixed at the beginning of system of season and remains throughout the season for short and frequent irrigation.
- **Permanent irrigation system:** The permanent system is suited to automation using soil moisture sensor and are generally preferred in orchard. The sub main, lateral are permanently buried below the ground level.

Based on Spraying Arrangement Fixed Head

The fixed head type of sprinkler arrangement system sprays water in one direction.

- **Rotating head:** Rotating head removes slow rate to distribute water in a circular fusion. They may be single, double, multiple nozzle sprinkler head. Single nozzle sprinkles system are referred for their low application rate however the double nozzle sprinkle head gives good uniformity of application at low pressure. Multiple nozzle type of sprinkler also called as giant sprinkler are used to covered more area with single set. The operating pressure required for such sprinkler may be more than 10kg/cm^2.
- **Perforated Pipes:** Such systems are preferred for application of water under lower operating pressure usually between 0.5 to 2.5 kg/cm2. The system is not recommended under heavy winds as the jets are distorted more easily. Generally, pipes are provided with holes, perforated along the upper 1/3rd perimeter in a proper designated to covered with between 6 to 15m such system are preferred on plains moderately high infiltration

rate used for irrigation of lawns or vegetable crop where the plant height ranges between 40 to 60 cm. Such systems are preferred for application of water under lower operating pressure usually between 0.5 to 2.5 kg/cm^2. The system is not recommended under heavy winds as the jets are distorted more easily. Generally, pipes are provided with holes, perforated along the upper 1/3rd perimeter in a proper designated to covered with between 6 to 15m such system are preferred on plains moderately high infiltration rate used for irrigation of lawns or vegetable crop where the plant height ranges between 40 to 60 cm.

Components of Sprinkler Irrigation System

1. ***Prime mover/pump suction pipe, foot value:*** Pumping sets or pump is required for lifting water from the source and push it through distribution system *i.e.* main, sub main, laterals and finally through the sprinkler head under sufficient pressure. The pumping set consists of a centrifugal pump (volume) or turbine type pump with a driving unit suction line and a foot value. Centrifugal pump are generally used where the lift is less than 5m *i.e.* when source is river, shallow well *etc.* for higher lift or if water level fluctuates widely turbine pumps are recommended. The electric motors are generally used driving unit for fixed installation diesel engines are generally recommended for portable units.
2. ***Main line:*** It carries water from the source (pumping unit) to the various parts in the field. It may fixed or portable. Permanent lines are generally buried below the working depth inside the ground. Light weight aluminum pipe with quick couples are preferred for portable lines something HDPF (high density pipes are also preferred because of its longer life. The fixed are generally of steel pipe or PVC pipe of suitable diameter). Te or L section is provided to connect the main with sub main or lateral.
3. ***Sub main:*** It carries water from main to lateral lines.
4. ***Lateral Lines:*** It carries water from main or sub main pipe line to the sprinkler head through the rise pipe. They are portable and equipped with quick coupling devices. Commonly they are available in 5,6 or 12m length are provided with U shaped rubber gasket in the female portion of coupling. The water pressure forces outside of 'U' gaskets to form water seal when the water is turned off the seal is broken and water is drained out from the pipe making it easy to uncouple and more.

Sprinkler Head

Sprinkler heads are used for spraying water on the fields they may be

a) Rotating Head

b) Fixed head Type and c) Perforated Type

Fixed Head Type: Used in landscape

Sprinkler lead can be classified on basis of pressure

1. Low operating pressure sprinkler (1.5 to 2.5kg/cm^2)
2. Intermediate pressure sprinkler (2.5 to 5kg/cm^2)
3. High pressure sprinkler (5 to 10kg/cm^2)

7.3.5 Drip Irrigation/Micro Irrigation

Micro irrigation is defined as the methods in which low volume of water is applied at low pressure and high frequency usually an irrigation interval is in the ranges of 1 to 4 days. The system has extensive network of pipes at operated at low pressure. At pre-determined spacing outlets are provided for emission water generally known as emitters.

In drip irrigation the required quantity of water is applied by means of mains, sub mains, manifolds and plastic laterals in the with equally spaced emitters usually laid on the ground surface at low pressure and at low discharge at the root zone of the crop.

Advantage of Drip Irrigation

- ☆ Water saving is up to 40 to 60 per cent
- ☆ Enhance the plant growth and increases the crop yield
- ☆ Savoring in level and energy most. Suitable for poor soil.
- ☆ Weed infestation is minimum
- ☆ Economy in cultural practices and easy operations.
- ☆ Chance of using saline water.
- ☆ Improve efficiency of fertilizers.
- ☆ Very flexible in operation
- ☆ No soil erosion.
- ☆ Easy installation, no land preparation.
- ☆ Minimizing quantity of produce.
- ☆ Enhances the maturity of the crop.

Limitations

- ☆ High maintenance requirement.
- ☆ Salinity hazard
- ☆ Economy limitations (40,000Rs/ha)
- ☆ High technical know-how is required.

7.3 Soil-Moisture Conservation

The availability of irrigation water in agriculture or in mulberry cultivation is becoming very much limited day-by-day due to gradual reduction in the ground water table. Hence, it is highly essential to conserve the soil moisture through different methods as described below for productive growth of crop plants.

7.3.1. Fall Ploughing

This is done by repeated ploughing of land just before the onset of monsoon for maximum percolation of rainwater.

7.3.2. Compartmental Bunding

Small sub-plots are made within the main plots with raised bunds to hold water in the sub plots.

7.3.3. Dead Furrow with Tide Ridges

This can be adopted in relatively sloppy lands. The furrows should be intercepted with small bunds to stop run-off water.

7.3.4. Growing Green Manure Crops

Application of sufficient organic matter, tank silt, and farmyard manures can be added to improve water-holding capacity of the soil. However, in the days of scarcity of these inputs, one can practise green manuring with leguminous plants like sun hemp, dhaincha, cowpea *etc.* to improve the organic matter content of the soil which in turn can conserve soil moisture effectively.

7.3.5.Mulching

Covering the surface of the soil with dry leaves, weeds *etc.* to check the loss of soil moisture through evaporation is called mulching. This can also be done economically by growing green manure crops and using the same as mulch material. The soil surface should be covered with dry weed mulch to check the soil moisture loss through evaporation. Under rainfed condition, the plants are entirely dependent on rainwater; therefore, it is very important to conserve soil moisture by means of mulching to avoid run-off loss of rainwater from the land.

7.3.6. Growing Drought Tolerant Mulberry Varieties

Some plants in nature can survive with limited water, which is governed by their genetic capacity to tolerate water scarcity and they are called drought tolerant plants. In mulberry also, there are varieties like S-13 and S-34, which are very efficient in withstanding drought conditions and can produce reasonably high leaf yield. These varieties grown under rainfed conditions can very well survive with efficient use and hence can be very useful in sericulture.

7.4 Soil Fertility, Organic Manure and Fertilizer Use

Mulberry is a fast growing plant and thus consumes the soil nutrients quickly. Regular manuring the mulberry field is required for the normal growth and maximum leaf-yield from the plants. Manures may be applied as organic or synthetic chemical forms or as biofertilisers.

Most frequently used organic manures include compost, sericulture wastes, poultry manure, green manure, oil cakes, biogas slurry, farm yard manure, *etc.* Organic manures increase the nutrient content of the soil, conserve soil moisture and enhance biolo¬gical activities of the soil.

Chemical fertilisers used for mulberry growth and yield are ammonium sulfate, ammonium chloride and nitrate, sodium nitrate, urea, potassium ammonium nitrate, bone dust, *etc.* Chemical fertiliser may be supplied to the plants along with irrigation water or directly as foliar sprays.

Biofertilisers suitable for mulberry cultivation are different nitrogen fixing bacteria like Azotobactor, Azospirillitm and aquatic fern-like Azolla. The Vesiculoarborescent mycorrhiza (VAM), Glomus fasciculatum are also grown in the mulberry field to enhance the PO4 utilisation by the plants. Bioferti¬lisers should not be mixed with chemical fertilizers unless specified.

Once planted, mulberry can give high yield of quality leaf consistently for more than 15 years. But this is possible only when the garden is maintained with proper intercultivation activities and soil fertility. Soil fertility means the capacity of the soil to supply essential nutrients for good growth of the plants.

Soils are highly variable with respect to their fertility status, physical and chemical properties. So, it becomes necessary for testing the soil periodically to know its fertility status and to take suitable corrective measures to maintain consistent high leaf yield and quality. The nutrient demand in case of mulberry is high due to intensive cultivation, harvesting of the leaves many times a year and also repeated pruning of the branches resulting in loss of nutrients from the soil rapidly. Therefore, it is very essential to replenish the deficient nutrients after assessing the soil fertility status.

Soil test should be undertaken at least once in two years., The following information will give an idea about soil fertility rating for important soil nutrients to take suitable corrective measures (Table 1).

7.4.1 Organic Manure

Organic manures are obtained from decaying plant and animal material. It plays an important role in building up soil fertility and increasing moisture-holding capacity and growth of micro-organism in the soil. It also improves micro-nutrient levels and the physical condition of the soil. Organic manures helps in improving the soil moisture retention in sandy soils and easy percolation of water in clay soils. Hence, application of adequate quantity of organic manure to maintain the fertility status of the soil is very essential.

Table 1: Fertility Rating of the Soil for Mulberry Cultivation

Sl.No.	*Nutrient*	*Fertility Rating*		
		Low (Below)	*Medium*	*High (Above)*
1.	Organic Carbon (per cent)	0-5	0.5-0.75	075
2.	Total Nitrogen (kg/ha)	250	250-500	500
3.	Available Phosphorus (kg/ha)	20	20-50	50
4.	Available Potassium (kg/ha)	125	125-300	300

The common organic manure used for mulberry garden is farmyard manure, Compost, poultry manure, press mud *etc.* Use of compost or vermicompost is very useful in increasing the nutrient availability to the plants. Under irrigated condition, farmyard manure is applied at the rate of 20 tonnes per hectare per year in 2-3 split doses, whereas, under rainfed condition, 10 tonnes of farmyard manure per hectare per year is applied during the onset of monsoon after annual bottom pruning. As the availability of good quality farm yard manure is becoming difficult day-by-day, it is highly beneficial for the farmers to prepare compost or vermicompost using sericultural farm residues like silkworm litter, rearing bed refuses, leftover mulberry twigs and other decomposable organic materials like weeds, leaves, green twigs *etc.*

Table 2: The NPK Content of some of the Common Organic Manure (per cent)

Manure	*Percentage Content*		
	Nitrogen (N)	*Phosphoric Acid (P_2O_5)*	*Potash (K_2O)*
Press mud	1-1.5	4-5	2-7
Animal refuse	0.3-0.4	0.1-0.2	0.1-0.3
Cattle dung, fresh	0.4-0.5	0.3-0.4	0.3-0.4
Horse dung,fresh	0.5 -0.5	0.4-0.6	0.3-1.0
Poultry manure, fresh	1.0-1.8	1.4-1.8	0.8-0.9
Sewage sludge, dry	2.0-3.5	1.0-5.0	0.2-0.5
Sewage sludge, activate dry	4.0-7.0	2.1-4.2	0.5-0.7
Cattle urine	0.9-1.2	Trace	0.5-1.0
Horse urine	1.2-1.5	Trace	1.3-1.5
Human urine	0.6-1.0	0.1-0.2	0.2-0.3
Sheep urine	1.5-1.7	Trace	1.8-2.0
Ash, coal	0.73	0.45	0.53
Ash, household	0.5-1.9	1.6-4.2	2.3-12.0
Ash, wood	0.1-0.2	0.8-5.9	1.5-36.0

Manure	Percentage Content		
	Nitrogen (N)	Phosphoric Acid (P_2O_5)	Potash (K_2O)
Rural compost,dry	0.5-1.0	0.4-0.8	0.8-1.2
Urban compost,dry	0.7-2.0	0.9-3.0	1.0-2.0
Farmyard manure,dry	0.4-1.5	0.3-0.9	0.3-1.9
Filter-press cake	1.0-1.5	4.0-5.0	2.0-7.0
Rice hulls	0.3-0.5	0.2-0.5	0.3-0.5
Groundnut husks	1.6-1.8	0.3-0.5	1.1-1.7

It has been estimated that one hectare of mulberry garden under irrigated conditions can provide about 6-8 metric tonnes of sericultural residues which can be used effectively for the production of compost/vermicompost.

7.4.2 Application of Chemical Fertilizers

The requirement of major nutrients especially nitrogen is very high compared to other crops as it is fast growing and grown mainly for its leaf which is harvested at least five times in a year. Mulberry is a nutrient hungry plant. Hence, application of recommended dose of chemical fertilizers is essential as fertilizers break down quickly and the nutrients are readily available to the plants for quick and luxuriant growth.

Chemical fertilizers are applied to the soil in the form of straight or complex fertilizers from time to time to improve the availability of major plant nutrients, *i.e.*, Nitrogen, Phosphorus and Potash. The quantity of fertilizers and the schedule of application mainly depend upon the system of cultivation. The recommended dose of chemical fertilizers and schedule of application for irrigated and rainfed mulberry in India is presented in Table 3.

Mulberry gardens maintained under irrigated condition with shoot cut method of harvest is provided with chemical fertilizer at the rate of 350:140:140 kg NPK/hectare/year in 5 equal split doses corresponding to five harvests at the rate of 70:28:28 kg NPK per hectare per crop. In mulberry garden under irrigated condition with leaf picking method of harvest, the fertilizer is applied at the rate of 300:120:120 kg NPK/ha/year in 5 splits. The first and third dose of fertilizer is applied at the rate 60:60:60 kg NPK/ha/crop in the form of straight or complex fertilizer while second, fourth and fifth dose of fertilizer is applied in the form of Nitrogen only.

Under rain fed condition, NPK is applied at the rate of 100:50:50 kg/hectare/year in two split doses. The first dose of NPK is applied at the rate of 50:50:50 kg/hectare after annual bottom pruning during rainy season (June-July). The next dose of 50 kg nitrogen alone is applied 2-3 weeks after second leaf harvest.

Table 3: Recommended Dose and Schedule of Chemical Fertilizer for Mulberry Garden Maintained Under Irrigated and Rainfed Conditions

NPK (kg/ha/year)	*Schedule*	
	Crop No.	*Quantity kg/ha/crop*
Irrigated garden with shoot harvest	I	NPK@70:28:28
NPK @ 350.140.140 in 5 equal splits	II	NPK@70:28:28
	III	NPK@70:28:28
	IV	NPK@70:28:28
	v	NPK@70:28:28
Irrigated garden and leaf picking method of harvest	I	NPK@60:60:60
	II	60: -: -
NPK@ 300:120:120 in 5 splits	III	NPK@60:60:60
	IV	60: -: -
	V	60: -: -
Rainfed garden and leaf picking method of harvest	I	NPK@50:50:50
	II	-
NPK@100:50:50 in 2 splits	III	50: - :-

Chemical fertilizers should be applied only after 20-25 days of pruning or leaf harvest. The most commonly used straight nitrogen fertilizers are urea (46 per cent N) and Ammonium Sulphate (20 per cent N), while for phosphorus and potassium, straight fertilizers used are Single Super Phosphate (16 per cent P_2O_5) and Muriate of Potash (60 per cent K_2O). Use of complex fertilizers like Suphala (15:15:15), Vijay complex (17:17:17) and Sampurna (19:19:19), for Nitrogen, Phosphorus and Potassium source respectively, is also very common.

Fertilizers are applied by opening a furrow of about 5-8 cm deep by the sides of the plants after 25 days of shoot harvest/leaf picking using bullock drawn plough followed by covering it with soil (Figure 1).

Figure 1: Application of Fertilizer.

Irrigation should be provided immediately after the application of fertilizer for better uptake of nutrients by the plants. However, to reduce the cost of cultivation and soil pollution, chemical fertilizers can also be supplemented through the use of biofertilizers.

7.4.3 Biofertilizer

Biofertilizers are made up of live or latent cells of microorganisms like bacteria and fungi capable of helping the plants to get nitrogen and phosphorus. Biofertilizers are used as a cost effective supplement to fertilizers and are also eco-friendly. In mulberry cultivation, application of Azotobacter biofertilizer called Seri-Azo (Figure 2) at the rate of 20 kg/ha/year in 5 equal split doses (4 kg/crop) corresponding to the application of 5 equal split doses of chemical nitrogen fertilizer has been recommended.

Figure 2: Biofertilizer "Seri-Azo".

Phosphate solubilizing biofertilizer called Seri-Phos has been recommended to be applied at the rate of 5 kg/ha/year in 5 equal split doses (1 kg/crop) corresponding to the 5 split applications of phosphorus fertilizer.

The biofertilizer Seri-Azo helps in nitrogen fixation in soil while Seri-Phos helps in dissolving insoluble phosphorus in soil and making it available to the plants. Both the biofertilizers are mixed with 200 kg of powdered farmyard manure and applied in between the rows of mulberry by opening furrows 10 days before chemical fertilizer application. By the application of biofertilizer, it is possible to reduce the application of both nitrogen and phosphorus fertilizers by 50 per cent. Similarly, in rainfed mulberry, Seri-Azo and Seri-Phos can also be applied at the rate of 10 kg and 1 kg/ha/year, respectively during rainy season in two equal splits by mixing with 200 kg of powdered farmyard manure.

Biofertilizer Product	*Bioagent Present in Biofertilizer Product*	*Dosage (Kg/ha/year) Recommended for Mulberry*
Seri-Azo	*Azotobacter chroococcum*	23
Seri-Phos	*Bacillus megaterium*	5
VAM	*Glomus mosseaeGlomus fasciculatum*	200 kg soil based inoculum

7.4.4 Green Manuring

Green manuring is an age-old practice of growing leguminous plants and incorporating the same into the soil to improve the soil fertility. Green manuring plants can fix atmospheric nitrogen in the soil by the action of root nodule bacteria (Rhizobium). In mulberry garden, green manuring (Figure 3) is done by growing short duration green manuring crops like Sun Hemp, Dhaincha, Cowpea and incorporating the green biomass along with the roots in the soil by ploughing. In a year, green manuring can be done once or twice during rainy season (June-July and September-October). About 25-30 kg seeds of a selected green manure crop is required per hectare each time to be broadcasted between the rows of mulberry plants after pruning and inter-cultivation operations. After 35-40 days of sowing or before flowering, the green plants are cut into small bits and incorporated into the soil by ploughing. This helps in increasing soil organic matter and also supplies additional nitrogen besides improving soil structure and water holding capacity of the soil.

Figure 3: Growing Green Manure Crop.

7.5 Mulberry Nutrient Management through Foliar Spray

The some macro and micronutrients, growth promoting molecules can spray on mulberry leaves through foliar spray. Maximum nutrient uptake is possible through foliar application of nutrients.

7.5.1 Use of Plant Growth Regulators in Mulberry

Plant growth regulators are organic compounds synthesized in specified plant parts in small quantity and are transported to the place of requirement leading to a change in physiological responses. Plant growth regulators can be classified into growth promoters and growth retardants. Plant growth regulators are auxins, gibberellin, cytokinin and growth retardants are Abscisic acid and ethylene. The latest one added to the growth promoter is Brassinosteroid, used to translocate the nitrogen and phosphorus. Triacontanol is one of the commercial formulations and used to increase the moisture and protein content of leaves, which ultimately built the disease resistance in silkworm. Plant growth promoting Rhizobacteria stimulates the plant growth regulators like auxins, gibberellins *etc.*, and help in better nutrient uptake and increase tolerance. Vermicompost also contains some plant growth regulators. The combined effect of different plant growth regulators will give positive result in mulberry growth.

The commercially available plant growth regulators are auxins, gibberellins, cytokinins, abscisic acid and ethylene. The newly included plant growth hormone as promoter is Brassinosteroid. Vermicompost and biofertilizers like Plant growth-promoting rhizobacteria (PGPR) are also having plant growth-promoting activity to increase the leaf yield and quality and also inducing the systemic resistance against the pathogen. Mulberry is the perennial crop and it is the sole food crop for silkworm, Bombyx mori L. Thus, the usage of plant growth regulators is also somewhat different. Under favorable conditions, plant growth promoters are increasing the sprouting of plants and enhance the leaf yield. Under unfavorable conditions, plant growth retardants are used to reduce the yield losses by reducing the evaporation and transpiration.

7.5.1.1 Auxins

Auxin is synthesis by Indole-3pyruvic acid (IPA) pathway by using the precursor Tryptophan. Plants largely produce auxin in shoot tips and translocate to roots. The primary auxin in plants is indole-3-acetic acid. The synthetic auxins are α-naphthaleneacetic acid (NAA) and 2,4- Dichlorophenoxyacetic acid (2,4-D). Auxin, mainly used to induce the apical dominance, fruit development and lateral root formation. Generally, auxin and auxin like compounds are inducing root formation, increase in root number and length. IAA at 500 ppm were increased the leaf yield by 67 per cent [Bose *et al.,* 1995]. IAA (Indole3-acetic acid) at 1000 ppm, improves leaf lobation and induces femaleness [Minimizawa, 1974].

7.5.1.2. Gibberellins

Regulators of Plant Height Gibberellins were isolated from the fungus Gibberella fujikuroi. Gibberellins are synthesized in apical portion of stems and roots. The main function is to stimulate stem growth through cell elongation and cell division, to promote seed germination and break the dormancy in seeds and buds. It slows the process of senescence (biological aging) by preventing the breakdown of chlorophyll in leaves. Suzuki and Kitano reported that GA stimulates shoot elongation but depresses the leaf enlargement and decrease the senescence. But GA found more effective under limited moisture conditions [Ohnishi,1986].

7.5.1.3 Cytokinin

Regulators of Cell Division Cytokinin, promote cell division, or cytokinesis, in plant roots and shoots. They are involved primarily in cell growth and differentiation, but also affect apical dominance, axillary bud growth, and leaf senescence. It largely produces in roots and translocates to shoot tip. Cytokinins represented by kinetin, zeatin and 6- Benzylaminopurine (BAP). Kinetin was the first cytokinin discovered and so named because of the compounds ability to promote cytokinesis (cell division). Though it is a natural compound, It is not made in plants, and is therefore usually considered a "synthetic" cytokinin (meaning that plant hormone is synthesized somewhere other than in a plant). The most common form of naturally occurring cytokinin in plants today is called zeatin, which was isolated from corn (Zea mays). Kinetin stimulates the root growth of mulberry cuttings and at 100 ppm increase the growth of auxillary buds [Lee, 1984]. The foliar spray of kinetin based PGRs (Biozyme) spray to mulberry plant before onset of water logging, showed that biozyme partially compensated the water logging effect and increased the leaf yield by 30 per cent and improved the chlorophyll, sugar content and photosynthetic rate significantly [Dhiraj and Kumar, 2011].

7.5.1.4 Abscisic Acid–Stress Hormone

The pathway for synthesis of ABA is Terpenoid pathway. It induce seed dormancy and desiccation tolerance activate the closing of stomata. Mulberry is hypostomata type, which is having stomata on lower surface and inhibits growth of plant. The use of ABA in tissue culture is limited because ABA suppresses the metabolism of callus [Ohnishi, 1986]. But it is useful for preservation of callus up to 20 weeks at 28°C. ABA levels increase positively with decreasing leaf water potential during drought condition. During stress condition synthesis of ABA inside the plant system cause to growth retardant and it reduce the respiration and evaporation. Ultimately it reduces the yield loss. Abiotic stress is known to modulate the content of abscisic acid in leaves of higher plants and this growth hormone is known to regulate many developmental events in plant growth including seed maturation, dormancy stress tolerance and water relations [Kolluru *et al.,* 2009].

7.5.1.5 Ethylene

Important role of ethylene is signaling the plants during onset of leaf senescence and abscission in deciduous plants and ripening the fruits. Ethrel at 400 ppm also induce femaleness in mulberry [Ogure,1980]. Ethrel at 0.1 ppm, induced the rooting in mulberry cuttings but not further root growth. Ethylene spray to axillary bud stimulates the onset of new shoot elongation at low concentration. Continuous application of ethylene in solution form found to depress the shoot elongation.

7.5.2 Micronutrient Based Foliar Spray

Foliar spray is a type of feeding technique to the plants by applying liquid fertilizers directly to their leaves. The absorption of essential nutrients takes place largely through the stomata of leaves and through the epidermis. Mulberry (Morus sp.) is a deep rooted high biomass producing foliage crop cultivated as a sole food for silkworm, *Bombyx mori* L. But due to repeated harvests and soil problems, mulberry is exhibiting nutrient deficiencies in recent years. Although crops need low amounts of micronutrients (Monreal *et al.,* 2015). Thus, there is a need to correct their deficiencies through use of foliar sprays. Furthermore, foliar sprays on mulberry plants help in disease management, enhance biochemical constituents of mulberry leaf, silkworm growth as well as cocoon parameters. The commercially used foliary spray in mulberry cultivation mentioned in the Tables 4 and 5.

Table 4: Some of Widely Used foliar nutrients in mulberry cultivation

Deficiency	*Corrective measure*	*Recommended Dose*
Manganese	Manganes sulphate ($MnSO_4$) by mixing with FYM and placed between the plant rows Spray of $MnSO_4$ to be taken 20–25 days after pruning. Generally 2–3 times application recommended with an interval of 4–5 days each	5--0 kg/ha/crop
Zinc	Foliar spray of 0.2 per cent of Zinc Sulphate ($ZnSO_4$) to be taken after 20-25 days after pruning	4.4 kg/ha/crop
Micro-nutrients	Foliar spray of Vipul/Triacontanol-cf (Plant growth regulators) Chemically Triacontanol (TRIA) is 30 carbon primary alcohol and it is present in cuticle of the leaf	250 ml/ha/harvest Spray given in two split of 125 ml each in 600 litres of water 1st split spray (125 ml/ha) given after 10-15 days after bottom pruning 2nd split spray (125 ml/ha) given after 15 days after first spray
Micro-nutrients	Foliar application of 0.25 per cent concentration of Poshan plus Seriboost	

7.6 Pruning

Pruning is one of the most important activities of mulberry cultivation. It is a process of removing certain branches systematically to give the mulberry plants a convenient shape and size and also to activate the buds to sprout and produce new branches.

Table 5: Commonly Used Foliar Nutrients in Mulberry Cultivation

Foliar Nutrients	*Improvement in Plant*	*Source of Information*	*Natural Product*	*Synthetic Product*	*Commercially available Product*
Botanicals	Chlorophyll-a, b, total chlorophyll, total soluble sugar, total soluble protein and total soluble phenol	CSR and TI, Berhampore, India.	+	–	–
B-nine	Leaf moisture and chlorophyll	RSRS, Dehradun, India.	+	–	–
Phalda	Leaf yield	RTRS, Bhandara, India.	+	–	–
Tracel, Navaras, Mircon 'S20' Amruth, Paras, Plantovit	Growth, quality and yield of mulberry leaves	DOS, Bangalore University, India.	+	–	–
Azotobacter choococcum	Atmospheric nitrogen	CSR and TI, Mysore, India.	+	–	–
Seri-Azo and Seri-Phos	Mulberry leaf yield and quality	DOS, UAS, Dharwad, India.	–	+	–
Di- ammonium phosphate	Total chlorophyll, crude protein, total carbohydrates, nitrogen phosphorus, potassium, calcium, magnesium and sulphur.	DOS, GKVK, Bangalore, India.	–	+	+
Daman Penshibao	Amino acids, citric acid and humic acid, photosynthesis	DOS, GKVK, Bangalore, India	+	–	+
Biozyme	Chlorophyll, sugar content and photo synthetic rate an water logging effect	CSR and TI, Berhampore, India.	+	–	–
VAM	Leaf quality and quantity	RSRS, Chamarajanagar, India.	+	–	+
Green Leaf	Shoot height, number of shoots, leaves per plant, fresh leaf yield, dry leaf yield, moisture content	DOS, Bangalore University, India.	+	–	–

Foliar Nutrients	*Improvement in Plant*	*Source of Information*	*Natural Product*	*Synthetic Product*	*Commercially available Product*
Seriboost	N, P and K constituents of the leaf	SDTCL, Bangalore, India.	+	–	+
Sodium, chloride	Soluble sugar and proline	CSR and TI, Berhampore, West Bengal, India.	–	+	+
Ethrel	Nitrogen content increased in dropped leaves	PPRI, DOS, Dokld-Giza, Egypt.	–	+	+
Plantonik	Chlorophylls, proteins, sugar and amino acid	RSRS, Jorhat, Assam, India.	+	–	–
Urea and NPK	Nitrogen in plants	BAAS, Dhaka, Bangladesh.	–	+	+
Plantoflex, Microflex	Growth, quality and yield of mulberry leaves	KSSR and DI, Bangalore, India.	+	–	–
Multizyme	Leaf moisture and chlorophyll	KSSR and DI, Bangalore, India.	+	–	–
Nostoc, Anabaena	Leaf yield	KSSR and DI, Bangalore, India.	+	–	–
2 Chloro ethyl-trim ethyl-ammonium chloride	Metabolic activities	CACNU, Kwangju, Korea Republic.	–	+	+
Harith	Leaf yield and quantity	KSSR and DI, Bangalore, India.	+	–	+
Tracel-2	Leaf spot disease	Rallis India, Ltd., Bangalore.	–	+	+
PGPR	Leaf rust disease	RSRS, Chamarajanagar, India.	+	–	–

+: Yes, –: No.

Pruning is periodical and methodical cutting of branches from mulberry plants. It is a very important operation in moriculture. It ensures:

- ✰ Good yield of quality leaves,
- ✰ Sprouting of young shoots from auxiliary buds,
- ✰ Good vegetative growth,
- ✰ Removal of dead and defunct wood, and
- ✰ Better exposure of the plant to sunlight, *etc.*

Pruning is done mainly to improve the yield and quality of the leaf. It is also followed to harvest the shoots for silkworm rearing and to carry out different inter-cultivation operations like ploughing, weeding, manure and fertilizer application and synchronization of silkworm rearing. Pruning is done using a sharp secateur, pruning saw or sickle without damaging the bark or stem.

Under rainfed conditions, the plants are pruned only once a year at 10-20 cm above the ground level during June-July at the onset of monsoon. Then, the plants are allowed to grow as bush, with many branches until the next year. Harvest of leaf is done by leaf picking method only.

7.6.1 Types of Pruning

Pruning is classified on the basis of the height of the stem at which it is cut. When the main stem is cut at ground level at every harvest without allowing for the formation of stump, it is called bottom pruning. In case of low-cut pruning, the plants are cut about 10-30 cm above the ground level for the development of the main stump. Medium cut and high cut pruning is followed at around 60 cm and 100 cm above ground for development of medium to small sized trees.

There are mainly two types of pruning:

i) Fist Form: Where the branches are cut at the base at the same height (Figure 4). Since no new branches are allowed to develop into new stump, it enlarges in width giving the shape of a fist to the plant.

Figure 4: Fist Form of Training.

ii) Non-fist Form: The branches are cut leaving 2-3 buds at the bottom of the shoot. Due to this, the height of the plants increases after every harvest of shoot. Non-fist form of pruning helps to produce more number of shoots (Figure 5).

Figure 5: Non-Fist Form of Stem.

7.6.2 Thinning of Branches

Thinning of branches is a process of removal of excess branches from a plant to facilitate it to grow well and to produce more leaves. When mulberry plants are pruned, many buds are sprouted and after a few days, you can observe that only few sprouted buds have developed into healthy shoots, while other sprouted buds remain stunted and weak. Such weak shoots are not suitable for silkworm rearing. In addition, the competition among many growing shoots causes improper distribution of plant nutrients among all the shoots and as a result weak shoots develop. Hence, it is very essential to restrict the number of branches (10-12) per plant by removing the excess branches after 30-35 days of pruning. This results in better growth, development of mulberry plants and production of quality leaves (Figure 6)

Figure 6: Thinning of Shoots.

7.7 Training

Regular pruning gives a specific shape to the mulberry plant which is referred to as training. Two types of training are given to mulberry plants:

- **(i) Fist Form Training**: In this type, the mulberry plant looks like closed fist when freshly pruned. Due to repeated pruning of branches at the same place every year, the height does not increase, but the diameter of the top part gradually increases. The plants may have single, two or even three fists.
- **(ii) Non-Fist Form Training:** In this case, the pruning is made at a level higher than the branching point every year. As a result, branching points of shoot increase in height every year not giving fist- shape to the growing plant.

Intercultivation

In all types of mulberry cultivation, the inter-row spaces are ploughed periodically either manually or by tillers. It helps to remove the weeds, loosen the soil with better aeration and percolation of water.

7.8 Weed Management

Weeds are unwanted and undesirable plants which interfere with the utilization of land and water resources and thus adversely affect human welfare. In broad terms, they are refered to as plants out of place. This implies that the weeds grow where we do not want to grow. However, that is crop in one country may be weed in another. Out of 3, 00,000 plants species, nearly 30,000 species behave as weeds. The weeds not only grow in cultivated lands but also grow in other places like water channels, industrial and urban sites, road sides, railway lines *etc.*

Table 6: List of Weed Plants Noticed in Mulberry Plantation

Sl.No.	*Botanical Name*	*Family*
1	*Achranthus aspera*	Amaranthaceae
2	*Amaranthus viridus*	Amaranthaceae
3	*Amaranthus spinosus*	Amaranthaceae
4	*Alternanthera sessilis*	Amaranthaceae
5	*Mikania cordata*	Asteraceae
6	*Ageratum conyzoides*	Asteraceae
7	*Solanum nigrum*	Asteraceae
8	*Cyperus rotundus*	Cyperaceae
9	*Cynodon dactylon*	Poaceae
10	*Euphorbia hirta*	Euphorbiaceae
11	*Leucas aspera*	Lamiaceae
12	*Mimosa pudica*	Fabaceae

Sl.No.	Botanical Name	Family
13	*Phyllanthus niruri*	Euphorbiaceae
14	*Tridox procumbens*	Asteraceae
15	*Eclipta alba*	Asteraceae
16	*Parthenium hysterophorus*	Asteraceae
17	*Physalis minima*	Solanaceae
18	*Xanthium indicum*	Asteraceae

7.8.1 Methods of Weed Control

7.8.1.1 Mechanical Method

Tillage: Tillage enables the removal of weeds from the soil It also buries the weeds. It is done by using implements drawn either by animal or by machines (tractor/tiller). Tillage is carried out both pre plant and post emergence. It facilitates deep burial of weed seeds during deep ploughing, leading to delayed germination. Shallow tillage helps the weed seeds to remain on the surface which germinate with the first rains and thus can be eliminated in young stage.

Hoeing: Rremoving weeds manually using "Hoe" It is effective method of weed control. The annual and biennial weeds with shallow root system can effective and completely control by this method.

Digging : It is effective against perennial weeds especially with deep root system and those propagating through underground parts like tubers, rhizomes, *etc.*

Mulching: Mulching which inhibits the emergence of weed and growth by exerting smothering effect. Mulching have the following advantages

- ☆ Suppress the weed growth
- ☆ Conserves soil moisture
- ☆ Increases the infiltration of water
- ☆ Reduces the soil temperature

7.8.1.2 Chemical Control

Employing chemicals for weed control constitutes chemical weed control. Herbicides are tool which enable man to grow desirable plants easily and cheaply. Best results with herbicides are obtained when they are used in combination with other agronomical practices including tillage. The use of herbicides have following advantages:

1. These can be applied for weed control in crop rows where cultivation is not possible.
2. It prevents the possible damage to roots.

3. It reduces the need for pre-planting tillage and is most effective in minimal tillage practices. Some herbicides used to control weeds are as under:
   - ✰ As post-emergence application, use Paraquat (Grammoxone) @ 2-3 lit/ha.
   - ✰ Spray Glycel 7.5 ml with 10 grams of ammonium sulphate per litre of water as post- emergence application. A total of 600 litres of spray fluid is required/ha.
   - ✰ Use flooding/deflector/fan type nozzle for spraying weedicide. Apply the weedicide immediately after pruning or within 2-3 days after pruning.

7.8.1.3 Biological Control

Biological control is generally defined as action of plants, predators or pathogens in maintaining another organism's population density at a lower average than would occur in their absence. While selecting a pest/predator/pathogen for biological, following aspects should be considered.

1. The biocontrol agent must ideally feed only on the target weed species.
2. Be free of its own predators and parasites and be resistant to those found in the areas introduced.
3. Be adapted to the local climate and have the capacity to seek out the host environment.

(a) Weed Control by Insects

The classical examples of control of some specific weeds by insect are: control of Goatweed (Hypericum *perforatum*) by the goatweed beetle (Chrysolina *hyperici* and C. *quadrigemina*).

Cactus (opuntia) by cactus moth (Cactoblastis *cactorum*) and cochinial insect (Dactylopius *indicus* and D. *tomentosus*)

Lantana *camera* by Crosidosema *Lantana*, Agromyza *lantanae etc.*

(b) Weed Control by Plant Pathogens

In recent years plant pathogens have become a good bio- control agent in controlling the problematic weeds. Some common examples are:

1. Use of *Alternaria macrospora* and *Puccinia heterospora* for the control of *Anoda cristata* in USA.
2. Use of *Cercospora rodmanii* for the control of water hyacinth, *Eichhornia crassipes.*
3. Use of the rust fungus *Puccinia romagnoliana* as a biocontrol agent for the control of purple nutsedge (*Cyperus rotundus*).
4. Use of the orange rust for the control of weed *Phyllanthus geniculate.*

7.9 Mulching

Sometimes, the inter row spaces of mulberry field are covered with cut-straw mats, or mulberry twigs or stubbles or leaf moulds, *etc.* It allows conservation of soil moisture, loose and fragile form of soil and also adds organic manure following their slow decomposition.

These above-mentioned steps, as formulated by research workers of different sericulture institutes for economic mulberry cultivation are generally followed by mulberry farmers. But some modifications in all steps according to local environment, types of silk moth and varieties of mulberry plant have been done by the field workers.

Green manure crops like cow pea, horse gram, or dhaincha can be grown in between mulberry rows during monsoon season and can be ploughed into the soil for improving the soil health and also serve as mulch

However, following the above package system of moriculture, yield of mulberry leaves has been raised to remarkable quality in both irrigated and rain-fed conditions.

Mulching with pruned mulberry twigs and other materials like straw and dried leaves will have the following advantages:

- Controls weed growth
- Conserves soil moisture by reducing run-off
- Increases the infiltration of water
- Reduces the soil temperature.

Chapter 8

Harvesting and Storage of Mulberry Leaves

8.1 Mulberry Leaves

The mulberry tree is native to Asia but is now found in many different countries across the world. Mulberry leaf has been used for centuries for its healing properties, culinary benefits, and incredible nutrition content. Traditionally, it was used to make tea or cooked and added to food. Mulberry leaves are caffeine and tannin free. It is now commonly found in dietary supplements.

The leaves contain six times more calcium than green tea, 25 times more than milk and 40 times more than cabbage. With respect to iron, it contains 2.5 times more than green tea and 10 times more than spinach.

One of the most amazing characteristics that can only be found in the mulberry leaf is an ingredient known as moranoline (1-Deoxynojirimycin/1-DNJ). This incredible compound inhibits an enzyme in the intestinal tract involved in the digestion of carbohydrates.

Moranoline or 1-DNJ holds back complex carbohydrates, starches, maltose and sucrose from breaking down into glucose. In essence, it actually prevents sugar from entering into the bloodstream. This is what makes it incredibly powerful for blood sugar.

The leaves also contain:

Table 1: Nutritional Value of Mulberry Leaves

Nutrients	*Content (per cent)*
Dry matter, per cent	89.30
Crude protein, per cent	29.80
Ether extract, per cent	5.57
Crude fiber	11.10
Gross energy, kcal/kg	4220
Ash, per cent	11.8
Neutral detergent fiber, per cent	35.80
Acid detergent fiber, per cent	28.00
Hemicellulose, per cent *	7.80
Calcium, per cent	2.73
Phosphorus, per cent	0.28
Amino acids	
Aspartic acid	3.06
Glutamic acid	3.33
Serine	1.22
Glycine	1.57
Histidine	0.69
Arginine	1.80
Threonine	1.36
Alanine	1.54
Proline	1.31
Tyrosine	0.82
Valine	1.77
Methionine	0.52
Cyctine	0.30
Isoleucine	1.43
Leucine	2.58
Phenylalanine	1.94
Lysine	1.88

8.2 Harvesting Mulberry Leaves

Harvesting is the collection of mulberry leaves from the plants in a rational method. It is an important part of moriculture that controls the growth of the silkworms in one hand, and quality as well as quantity of cocoons spun by them on the other. Like moriculture, harvesting also follows some general practices.

8.2.1 Time of Harvesting

Once the leaves are harvest from mulberry plant, the freshness goes on decreasing. In view of the active photosynthesis and transpiration during the day time, the leaves harvested late in the afternoon contain less water and more carbohydrates. The leaves harvested in the afternoon wither more rapidly.

Therefore, mulberry leaf should be harvested in the early morning, as the day passes the quality of leaves in terms of moisture content comes down due to active transpiration. Hence, leaves are picked up during the morning or evening hours and preserved carefully.

Harvesting of mulberry under row system of cultivation comprises the leaves grown on primary shoots, which are vigorous in growth and thin. The thinner leaves trend to dry faster. One of the reasons for thin leaves is close plantation. The alternate method to keep the quality of such leaves is feeding the worm by twig feeding or cutting the twigs along with leaves. One of the problems in this type is that the leaves grow faster on thinner branch; they have a tendency to go over-matured and turn yellow. During rainy season the moisture content will be more and increases the humidity in rearing bed.

When there is high humidity in rearing beds the outbreak of diseases like muscardine and grasserie takes place.

Hence, following care should be taken.

1. Avoid over feeding.
2. Providing proper ventilation in rearing rooms.
3. Harvesting leaves only when fully mature.
4. Proper bed cleaning.
5. Allow the bed to dry up fully.

In case of irrigated mulberry under pit system the leaves will be thicker and heavier, contain less moisture content during rainy seasons. One of the important to point to keep in view is, one irrigation should be given to the garden before feeding the leaves to the worms after the 4th moult. So that, succulent leaves are available for the silkworms of last instars.

Leaf picking should be done during early hours of morning or evening. For chawki worms tender leaves *i.e.,* 2nd and 3rd leaves below the growing bud are picked. In case of 3rd and 4th stage worms 3rd and 4th leaves are picked and as the stage of worms advances mature leaves can be fed

8.2.2. Choice of Leaves

From the quality point of view the requirements of young worms are completely different from those of late age worms. Mulberry leaves for young silkworms must be soft and rich in water content (78-80 per cent), protein (26-28 per cent) and carbohydrates (11 per cent). The top tender leaves meet these requirements.

The quality of leaf may vary considerably from season to season. In summer the leaves grow and mature fast, but wither quickly. Its life in the rearing bed becomes short favoring to increase number of feeds per day. In rainy season, the leaves grow and mature fast and contain more moisture. Hence the life is longer therefore reduces the number of feeds per day. Because of high moisture content, humidity of the beds increases. Therefore it is necessary to keep down the humidity of the bed under control through feeding of reduced quantity of leaf and more nature leaf containing less moisture. In cold season the mulberry growth is slower and leaves mature gradually. These leaves have better quality, optimum moisture and better feed value. This type of leaves neither withers nor increase bed humidity leading to successful crops. Morning time is the proper time for picking the mulberry leaves, planning the total leaf requirements of the day.

8.2.3 Leaf Selection Methods

Mulberry for young silkworms has a great effect on the growth and health of silkworm. Therefore selection of leaves should be done carefully. The leaves for silkworms must be soft, tender, rich in water content, protein, carbohydrates *etc.* There is high correlation between moisture content in the top tender leaves and chawki worm growth and moulting. Therefore for plucking the correct leaves for young worms, the largest glossy leaf method is adopted.

1. Glossy Leaf Method

For the identification of glossy leaf, hold the upper part of shoot lightly, and move the hand upward gently. Then appears a large leaf which stands out at the top (Figure 1). This is identified as largest glossy leaf.

Otherwise holding the upper part of the shoot lightly between fingers, and bend it horizontally. Then a leaf stands up right is identified as largest glossy leaf (Fig.2).

Now the first instar are fed with 4th and 5th leaves downward from the base of glossy leaf, for second instar 5th - 7th leaves, for third instar 7-8 leaves to downward are plucked.

2. Lenticel and Bud (LB) Method

In this method the colour of the lenticel and auxiliary buds are used to harvest mulberry leaves corresponding to the stage of worms. The colour of auxiliary buds changes from the top to the bottom of the branch. The colours from top to bottom are green, apical brownish, striped, non-accomplished and accomplished buds (Figure 2)

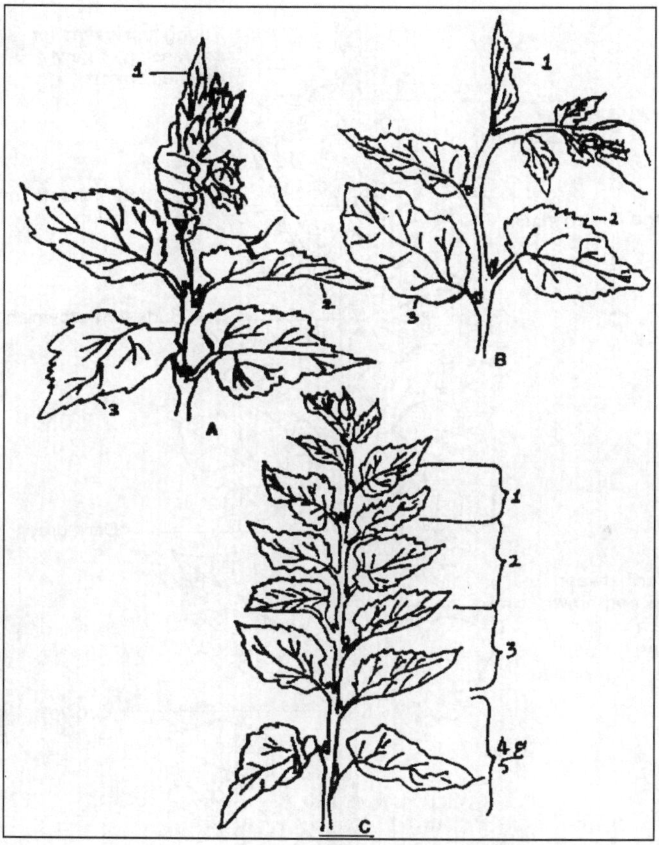

Figure 1: Glossy Leaf Selection (A,B,C).

Leaf with yellowish lenticel at the base of leaf petiole found above the apical brownish bud is used for first instar. Leaves from the brownish lenticel through the apical brownish bud are for second instar. The leaves from the brownish lenticel through the non-accomplished bud are used for third instar.

Mulberry Leaves for Warm-Feed

However, in every sericulture practice, following leaves are to be avoided for harvesting or if harvested, not to be fed to the larvae: decline leaves, buds, too tender or too mature leaves, pest infested or diseased leaves, insecticide sprayed leaves and too coarse leaves.

8.2.4 Equipment and Tools Harvest and Postharvest Management

Equipment and tools in direct contact with mulberry leaves should be made of non-contaminated materials. Knives or scissors shall be sharp and clean.

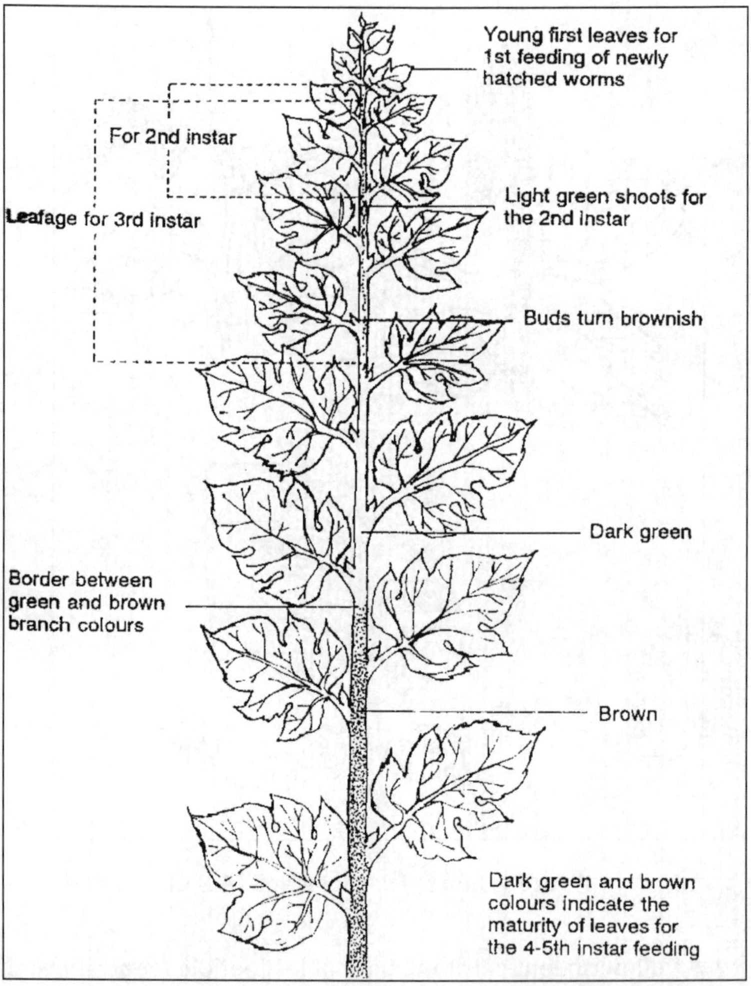

Figure 2: Lenticel and Bud Method.

Containers for transferring mulberry leaves shall be clean and not affect mulberry leaf quality. The harvested mulberry leaves shall not be placed directly on the floor.

The containers for harvesting and transferring shall be clearly separated from the containers used for refuses and pesticides so as to prevent contamination.

8.2.5 Harvested Mulberry Leaves for Silkworm Rearing by Silkworm Ages

The young silkworm, pick the first to the third unfurled mulberry leaves from the top shoot to feed as the first meal and after that pick the next top-down leaves.

The grown silkworm, cut the whole shoot or pick the mature leaves to feed grown silkworm except the top leaves.

8.2.6 Collecting, Handling and Transporting

Collecting area and vehicles for transportation shall be maintained as follows:

- ✰ The collecting place of mulberry leaves should be clean, and hygienic such as distanced from waste disposal area and the storage for chemicals, chemical fertilizer, and organic fertilizer. The place shall be of good air flow, without heat accumulation, able to prevent the leaves from disease carrier animal and other pets.
- ✰ Containers for packing mulberry leaves shall be separated from containers used in transferring or transporting pesticides or fertilizer to prevent chemical, biological, and physical contaminations which are harmful to consumption and causing damage to leaf quality.
- ✰ Vehicles for mulberry leaf transportation shall be cleaned prior to transporting. It is prohibited to transport mulberry leaves in the same load with hazardous substances in order to prevent contamination.
- ✰ Transporting shall be operated with care. The practices shall not cause damage or deterioration or harmful contamination.

8.2.7 Methods of Harvesting of Mulberry leaves

Leaf Picking

In this case, individual leaf with petiole is picked from the main branch. The terminal buds are nipped off at the same time. It helps the lateral shoots to grow rapidly so that sufficient leaves are available during next harvesting. After 10 weeks of bottom pruning, leaf picking can be started with intervals of 7-8 weeks for subsequent pickings. By leaf-picking desirable leaves can be collected but this method requires more labour, time and further, the leaves may get withered too quickly. This method is suitable for small scale rearing operation.

Advantages

1. Wastage of leaf can be minimized.
2. The main advantage of this method isthat, leaves can be selected to suit the growth stage of larvae.

Disadvantages

1. It requires more labour.
2. Leaves wither too quickly if not stored properly.

Whole-Shoot Harvest

In this method, branches/shoots are cut to the ground level so the sprouting shoots can reach uniform maturity by the next harvest. The entire shoots collected are given to mature larvae (after IV moult). Whole shoot harvesting is done either manually or by machine at an interval of 10-12 weeks. It is practised in Karnataka, West Bengal in India, Japan and Russia also.

Branch Cutting

Here, the entire branch is harvested and used to feed worms after III moult directly. It is applicable for tree and middling plantations. This harvesting method saves time and labour in harvesting and bed cleaning operations also. Further, leaves of entire branch remain fresh for a longer period. It is practised in Kashmir, West Bengal, parts of Karnataka in India and also in Japan and Russia.

Advantages

1. Cost of labor can be minimized.
2. Easy to feed the silkworm.
3. In this type of feeding wastage of leaf can be minimized.
4. Bed cleaning process is easy and labor saving.
4. Quality of leaves can be maintained for a longer duration, since leaves are not detached from the branch, and the succulence will last longer.
5. Easy to disinfect the rearing equipment if branch feeding is adopted in shelf rearing or floor rearing, since much of the equipment like rearing trays, feeding stand *etc.*, is not required.
6. Easy to preserve and maintain quality.
7. Hygienic conditions can be maintained effectively in rearing house.
8. Maximum utilization of mulberry leaves by silkworm.
9. Saves labor in collection of leaf, disinfection of feeds, bed clearing and spacing.

8.3 Preservation of Harvested Leaves

Harvested leaves are usually not given immediately for feeding to the larvae because of greater starch content which is not palatable. Moreover, the larvae feed 4-5 times a day while leaves are harvested 1-2 times per day. A certain time is needed to reach the harvested leaves to rearing place. For these reasons, mulberry leaves are needed to be preserved after each harvest.

During storage, the leaf loses water by transpiration and hydrolysis. Water loss is directly proportional to storage time. During storage, nutritive value of the leaves also deteriorates.

- ✰ To overcome these problems, harvested leaves may be stored or preserved by following ways:

- ☆ Leaves to be fed to young larvae may be kept in baskets but for grown-up larvae can be stored in wooden or bamboo leaf chambers covered on all sides with wet gunny cloth.
- ☆ For better preservation, leaves may be kept in gunny cloth bags of 22† x 32† size, which are double layered and easy to handle.
- ☆ Leaves should be kept in cool place.
- ☆ Exposure to air current should be avoided.
- ☆ Heap storage must not be done.
- ☆ The storage time must not exceed 24 hours.
- ☆ If leaves get rotten or dry, must be discarded.

Chapter 9

Biology of *Bombyx mori*

9.1 Introduction

In India, silk production percentage was 15 per cent of total world production. Generally, four kinds of silkworm are used for silk production. Those are Mulberry, Eri, Tasar and Muga. But in these Mulberry silkworms are producing 89.45 per cent of silk [1]. Production of silk is major agro-based rural industry. This Bombyx mori L. is a monophagous insect that feeds exclusively on mulberry leaves. In India, five states *i.e.* Karnataka, Jammu and Kashmir, Tamilnadu, Andhra Pradesh, and West Bengal are the traditional state for silk production.

The mulberry silk moth, Bombyx mori belongs to the family Bombycidae. China is the native place of this moth, but now it is totally domesticated and successfully reared in India, Japan, Korea, Italy, France and Russia.

In India 92 per cent of country's silk production comes from mulberry moth. The worm or larva of this moth solely feeds on leaves of mulberry plant, so named as, mulberry worm. Its wild ancestor B. mandarina still occurs in some isolated areas of Himalaya's foothill.

9.2 Taxonomy of *Bombyx mori*

Scientific Classification

Kingdom: Animalia

Phylum: Arthropoda

Class: Insecta

Order: Lepidoptera

Family: Bombycidae

Genus: Bombyx

Species: mori

Binomial name: *Bombyx mori* (Linnaeus, 1758)

9.3 Life Cycle of *Bombyx mori* L.

Bombyx mori being a holometabolous insect, undergoes complete metamorphosis with four stages in its life cycle, *viz.*, egg, larva, pupa and adult. Sexual dimorphism is present in adult, pupal and larval stages of life cycle.

The duration of each stage varies according to inherited characters of the race and also with the available environmental conditions, quality of food provided, *etc.* Here, the life cycle of most prevailed race of Bombyx mori in India will be discussed with the following headings.

9.3.1 Adult Moth

The adult silk moths do not take any food and incapable to fly. Their life-span is only 3-6 days. The whole body along with wings remains covered with epidermal scales and the body is divisible into 3 parts, *viz.*, head, thorax and abdomen.

- ☆ **Head:** The small head bears mouth at ventral side and two prominent bipectinate and serrate antennae on two sides. Below the base of antennae, two compound eyes are present on either side of the head. Mouthparts are located between the compound eyes, which include inactive coiled proboscis.
- ☆ ***Thorax:*** Thorax has 3 segments – prothorax, mesothorax and metathorax of which mesothorax is the largest. Each thoracic segment bears a pair of 5-jointed legs. The meso- and metathorax bear two pairs of wings, the larger, triangular forewings remain attached to mesothorax while smaller hind-wings are attached to the metathorax. Thorax also bears spiracles.
- ☆ ***Abdomen:*** The male moth has eight abdominal segments with a pair of hooks 'harps' at the caudal end which help in copulation. The female has larger and fatter abdomen than that of the male which has seven segments. A knob like projection covered with sensory hairs at the caudal end of female is present. It can secrete pheromone.

9.3.2 Copulation

Following emergence from the cocoon, the females rest on the cocoon, spreading out her feeble wings and immediately begin to scent by exuding pheromone from her abdominal projection. The males being attracted by the pheromone find the female and start mating (Figure 1).

Figure 1: Copulation between Adult Male and Female Silkmoth.

9.3.3 Egg Laying

After mating for 12-24 hours, the pair gets separated and the female starts laying eggs while the male moths die. Most females lay 350 - 400 eggs over the course of 1-2 day.

According to the race, *B. mori* can lay two types of eggs-diapausing or hibernating and non-diapausing or non-hibernating eggs.

9.3.4 Egg

The clustered eggs always remain covered with gelatinous secretion of the female moth. The colour of eggs depends on moth-race. In diapausing eggs, the colour changes from light yellow to deep brown when they enter diapause.

In non-diapausing races, the colour of eggs does not change till the development is complete. The eggs are ovoid, spherical or ellipsoid in shape and are flat on one side (dimple) through what eggs remain attached to the substratum.

Duration of embryonic development varies from 9-14days depending upon the race, season and nutrition during the larval stage of the mother. Duration of the life cycle spent in the egg stage depends on whether it is a diapausing or non-diapausing egg.

Table 1: Duration of Developmental Stages of Uni-bi-and Multivoltine Races of *B. mori* L.

Stage	*Uni/Bivoltine Races (Days)*	*Multivoltine Races (Days)*
Embryonic	11-14	9-12
Larval	24-26	20-24
Pupal	12-15	10-12
Adult	6-12	3-4

9.3.5 Hatching

The hatching time of silk moth eggs varies with voltinism and environmental conditions. Diapausing eggs under natural conditions remain dormant for months till next spring. If diapause is broken by acid treatment, the eggs have to be incubated at a temperature of 23-25°C for 11 to 14 days for completion of embryonic development and the larvae to hatch out. Non-diapausing eggs normally take 9-12 days to hatch into larvae.

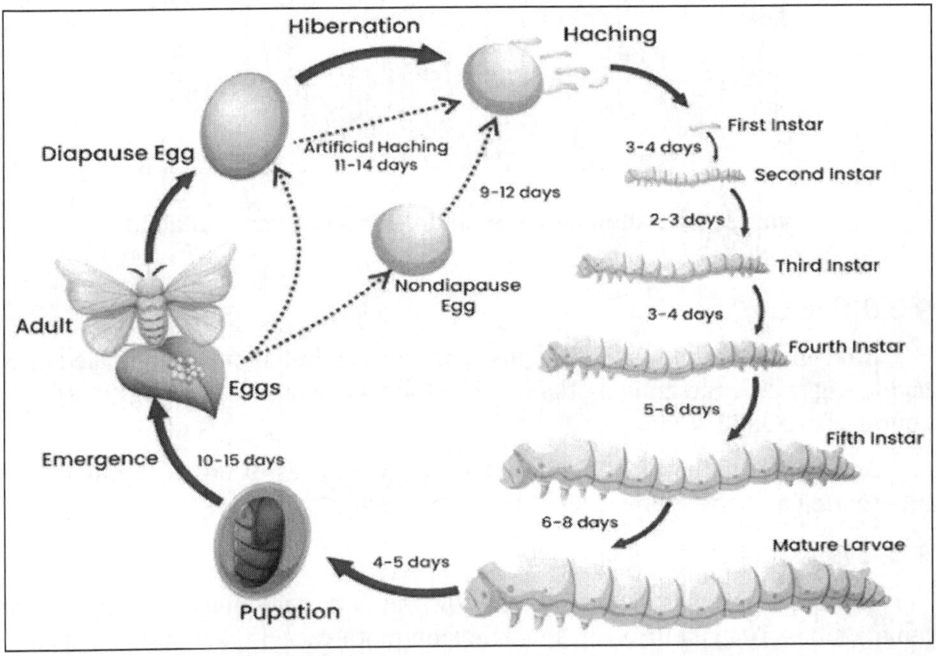

Figure 2: Life Cycle of Mulberry Silkworm, *Bombyx mori* L.

9.3.6 Larvae of *Bombyx mori*

The newly hatched larvae are about 3 mm long, black in colour and is covered by bristles situated on tubercles in each body segment. Young larvae soon after hatching start feeding on mulberry leaves and grow. This period is called first instar.

After about 3 days, larvae stop feeding and undergo ecdysis (shading off old cuticle). This period of ecdysis is called moulting. After moulting, the larvae start feeding again and grow rapidly in size. They repeat the moulting process usually for 4 times and thus the larvae or caterpillars get changed into second, third, fourth and fifth instar stages during the whole larval period.

It takes about 21-25 days after hatching in case of multivoltine moths or 24-48 days in case of uni- or bivoltine moths. The fifth larval stage is the longest stage when the larvae show maximum consumption of mulberry leaves and high growth.

Of the total ingestion during the entire larval period, nearly 85 per cent is taken during 4th and 5th in stars and a larva can take mulberry leaves weighing 30,000 times of its body weight to reach 5th instar stage from 1st instar.

The full-grown 5th instar larvae are 7.5-10 cm in length. Its body is divisible into 3 parts, small head with special spinneret through which silk is exuded; thorax with 3 segments, each with a pair of clawed true legs which are not used for walking but for holding the leaves while feeding; and abdomen with 11 segments.

The 9th, 10th and 11th abdominal segments are fused together to form the apparent 9th segment, anal plate and paired caudal prolegs. Each 3rd, 4th, 5th and 6th segment bears a pair of fleshy, un-jointed muscular protuberances, known as pseudolegs or prolegs, each bearing curved hook on the outer edge of leg tip disc. The eighth abdominal segment has a projection on dorsal side, called caudal horn (Figure 2).

The most characteristic feature of silkworm is their silk gland which is well-differentiated in 4th and 5th instar stages (discussed elsewhere). There are nine pairs of spiracles or stigmas placed laterally on either side of the larval body. Besides this, larva has a number of projections or nodules all over the body and markings on the skin which is race-specific. The larvae have sexual dimorphism. In female larvae, the sexual marking appears as a pair of milky white spots known as Ishiwatas' gland, one each in the right and left sides of 8,h and 9,h abdominal segments.

In male larvae, the reproductive gland appears as a small milky white protuberance at the centre of the ventral side between 8th and 9th abdominal segments (Figures 3 and 4). This gland is known as Herold's gland.

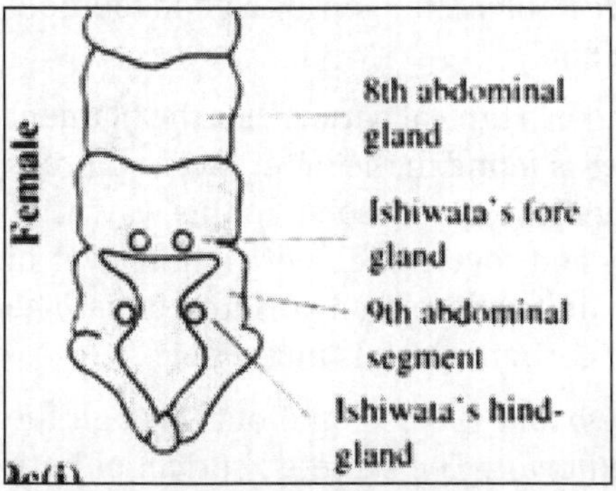

Figure 3: Male Reproductive Gland (Herold's gland).

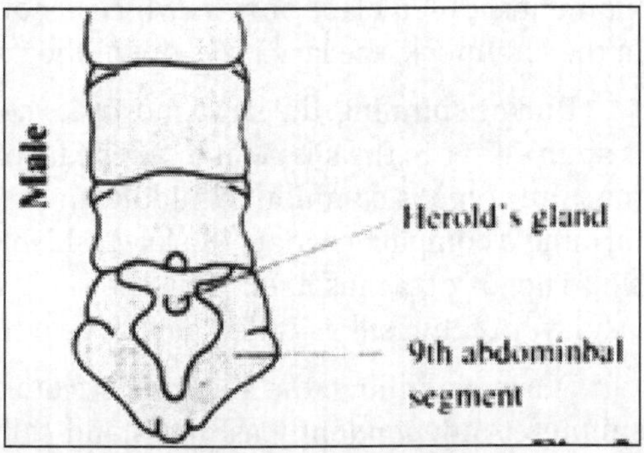

Figure 4: Female Reproductive Gland (Ishiwatas' gland).

9.3.7 Ripe Larva

The full-grown final 5th instar that has stopped feeding and ready to spin the cocoon is called ripe worm or ripe larva. This stage is attained after 6-7 days of feeding in 5th instar stage for bi- and multivoltine races in tropics and 7-9 days for uni- and bivoltine for subtropical races.

It is translucent as the gut discards all of its green content. The body of ripe worm shrinks in length and a visible constriction appears at 4th and 5th segments. The worm stops feeding and usually moves towards the edge of rearing tray.

Figure 5: A Photograph of Ripen Worm (*B. mori*).

9.3.8 Cocooning/Spinning

The ripe worm after being shifted to mountage passes last excreta which are red in colour due to presence of tryptophan metabolites. After emptying its gut, it first secretes a tiny droplet of silk – the anchorage spot, which hardens and sticks onto the mountage. The larva then anchors itself to that spot and spins a loose hammock forming the framework of the cocoon. This loose, tangled mass of filament is called blaze or floss which is not reelable. After getting a good foothold in the hammock, the larva lays down the cocoon.

During spinning, the larva moves its head in a typical pattern; so, the filament is spun either in the shape of or a. The former is found in the outer layer, whereas the latter type is common in middle and inner layers of cocoon. In this way, after forming a compact cocoon, the larva shrinks and covers itself with a thin layer of silk. Then it detaches itself from the shell, moults and transforms into pupa. The last layer of thin silk is called pelade layer of cocoon which is unreelable.

During spinning, the secretions from two silk glands come out through the spinneret independently as fibres and called brins. The secreted sericin of both glands cements these two brins in a single thread, called bave. Dragging – a physical process is also responsible for the coagulation of liquid silk to the fibre. The process of spinning the cocoon takes about one to two days in case of multivoltine and two to three days in uni/bivoltine races.

9.3.10 Pupa of *Bombyx mori*

Once the cocoon formation is over, the larvae moult inside and transform into pupae. This is known as pupation. Papal stage is also called as resting or inactive stage in the life cycle of silk moth. Actually, it is a transitional stage during which some changes take place in the body of mature larvae (ripe worms).

Before transforming into proper pupa, the mature larvae pass through a pre-pupal stage during which dissolution (histolysis) of many larval organs like silk gland, moulting gland, abdominal appendages take place. It is followed by the formation of adult organs (histogenesis) during pupal stage when a pair of large compound eyes, a pair of large antennae, fore and hind wings, legs, *etc.* become prominent.

The cuticle of newly formed pupa is soft and pale yellow in colour which gradually becomes darker. In the female pupa, a fine longitudinal line joining the 8th and 9th abdominal segments is present on the ventral side. In male pupae such line is absent; instead, a small opening is present in the 9th segment (Figure 6). Usually, female pupae are larger and heavier than the male pupae.

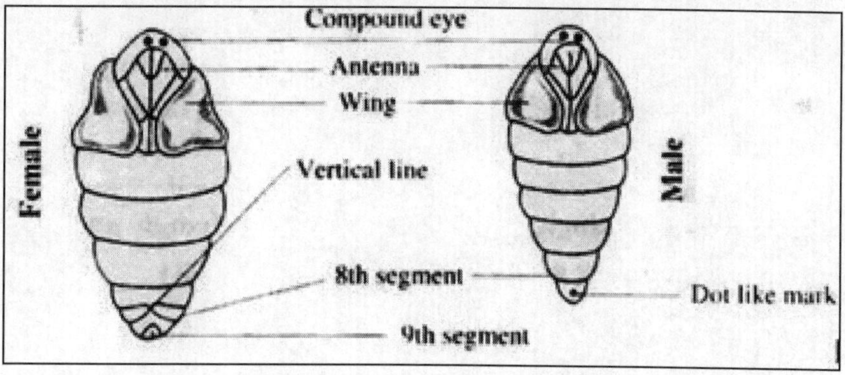

Figure 6: Male and Female Pupa of *Bombyx mori*.

The pupal stage continues about 10-12 days during which development of full-grown moth is completed. The adult moth then secretes mild protease from their salivary gland to dissolve the cocoon membrane in the form of a small opening through which they emerge out. In this way Bombyx mori completes its life cycle through the above-mentioned stages and concomitant activities.

Table 2: Duration of different Stages in the Life Cycle of Mulberry Silkworm

Stage	*Multivoltine*	*Uni-bivoltine race*
Egg	9-12 Days	10-14 days
Larva	20-24 Days	22-26 Days
Pupa	10-12 Days	12-15 days
Adult	3-4 days	6-8 days
Total life cycle	42-52 days	50-63 days

Chapter 10

Morphology of Mulberry Silkworm *Bomyx mori* L.

10.1 Introduction

Sericulture is an art of nurturing silkworm for the production of cocoon which is the raw material for the production of silk. Silk is the most cherished of all textiles even today, along with a wide variety of man-made fibre of inimitable excellence as "Queen of Textiles".

10.2 Morphology

10.2.1 External Morphology of Silk Moth

- An adult silk moth is strongly built, about 2.5 cm long and is creamy white in color.
- The body is divided into head, thorax and abdomen.
- The head bears a pair of compound eyes and a pair of feathery antennae.
- The proboscis is reduced, so they don't feed.
- The thorax bears three pairs of legs and two pairs of wings.
- Minute powdery scales cover the entire body.
- Male and female are distinct where the female is bigger than the male with distended body.

10.2.2 Egg

The size, weight, shape, color of the egg as well as number per laying vary among different strains. The silkworm eggs are tiny and weight of a single egg is about 0.55 to 0.6 mg, so that there are 2000 eggs per gram. The silkworm eggs measure 1 to 1.3 mm in length and 0.9 to 1.2 mm in width.

- ✰ The egg is covered by an egg shell called chorion which is provided with follicular imprints on surfaces. At the anterior end of egg there is micropyle.
- ✰ The egg contains a good amount of yolk and is covered by a smooth hard chitinous shell. After laying the eggs, the female moth does not take any food and dies within 4-5 days.
- ✰ In the univoltine (a single brood per year) they may take months 7 because over-wintering takes a place in this stage but the multivoltine broods come out after 10-12 days. From the egg, mulberry leaf hatches out a larva called the caterpillar

Female lays about 350-400 small, white, round (pin head shaped) eggs in cluster on mulberry leaves and she dies.

In cold, the eggs don't hatch for a long time and the increase in temperature causes them to hatch faster.

In sericulture, the eggs are collected and stored in a cold place when the food (mulberry leaves) are not abundant to feed the larvae. In the early spring, with the enough supply of food, the eggs are hatched inside the incubator at the temperature of 18-25$^{o.}$ Eggs hatch into larvae within 10-12 days.

10.2.3 Larva

The silkworm larva when newly hatched is black or dark brown. It has a large head and the body is densely covered with bristles so that it looks like a hairy caterpillar. There are four pairs of tubercles; the subdorsal, supra spiracular, infra spiracular and basal tubercle each of which carries three to six setae.

As the larva grows, it becomes smoother and lighter in colour due to the rapid stretching of the cuticular skin during the different instars of the larval stage. The larval body is composed of the same segments as any insect: head, thorax and abdomen.

The entire body is covered with a thin and elastic chitinous cuticle which is capable of being extended considerably to permit rapid growth of the larva during any instar. The integument consists of the epidermis and the cuticle over it.

- ✰ The newly hatched caterpillar is 6mm long, brownish in color and moves in a characteristic looping manner.
- ✰ The body is divided into head, thorax and abdomen.

- ✰ They are actively voracious and feed on mulberry leaves for 25-32 days.
- ✰ They grow very fast and undergo four moltings (changing the skin) to change into different instars.

10.2.4 Head

The head consist of six body segments fused together with a cranium. The second, fourth, fifth and sixth segments carry appendages which are modified into antennae, mandibles, maxillae and labium respectively. Median epicranial suture is well developed and prominent. Similarly, on the outer side, the clypeus and the labrum are also prominent.

There are six pairs of ocelli or larval eyes which are located behind and a little above the base of the antennae. There is a pair of antenna segments and these are used as sensory organs (feelers). The mandibles are well developed and powerful and are adapted for mastication.

Mouth parts are adopted for biting and chewing (mandibulate type) represented by two hard powerful mandibles for cutting or crushing the food. Two maxilla with 2 or 3 segmented maxillary palpi. Submentium is divided into 3 sclerites, mentum very large and less chitinised part. Prementum is born on mentum carrying spinneret and 2 segmented labial palp on either side

10.2.5 Thorax

The thorax consists of three body segments called the pro, meso and metathorax. Each of the three thoracic segments carries ventrally a pair of legs each comprising in turn three jointed segments.

These are the true legs which are conical in shape and carry sharp distal claws. These claws are not used for crawling but for holding mulberry leaves while feeding. All the silkworm larvae that show body larval markings carry the so-called eye spot on the dorsal side of the mesothorax.

10.2.6 Abdomen

The abdomen is comprised of eleven body segments although only nine can be distinguished as the last three are fused together to form the apparent ninth segment, the anal plate and the caudal legs. The third to sixth and the last abdominal segments each bear a pair of abdominal legs which are fleshy, unjointed muscular protuberances. At the extremity they form a sort of disc with a series of hooks inwardly curved and arranged in a semicircle. On the dorsal side of the eighth abdominal segment, the larva carries the caudal horn. The abdominal segments carry the sexual markings which develop distinctly in the fourth and fifth instars in the eighth and ninth segments on the ventral side. In the female the sexual markings appear as a pair of milky white spots in each of the eighth and ninth segments.

The pair of spots on the eight segments is known as Ishiwata's fore glands and the pair on the ninth segment is referred as Ishiwata's hind glands. In the male a small milky white body (Herold's gland) appears at the center of the ventral side between the eighth and ninth segments. There are nine pairs of spiracles placed laterally on either side of the body. They are found on the first thoracic segment and the first eight abdominal segments. These are the breathing or respiration pores.

The larval skin or integument consists of the cuticle and the hypodermis. The cuticle is made of chitin as well as protein and is covered with a thin layer of wax. Nodules are found all over the body surface of the silkworm larva. The distribution of these nodules differs according to the variety of silkworms, being less in number in varieties having lustrous body surfaces, such as the Chinese bivoltine and the multivoltine varieties. The nodules are found in larger numbers over parts of the body where normally the larval markings are found. The larval markings in silkworms are caused by skin pigment.

10.2.7 Pup

The pupal stage is generally called the resting, inactive stage of the silkworm when it is incapable of feeding and appears quiescent. This is a misnomer. The pupal stage is a transitional phase during which definite changes take place. During this period of biological activity, the larval body and its internal organs undergo a complete change and assume the new form of the adult moth. The mature silkworm larva passes through a short transitory stage of pre-pupa before becoming a pupa. During the pre-pupal stage, the dissolution of the larval organs takes place and this is followed by the formation of adult organs during the pupal stage. Soon after pupation the pupa is white in color and soft but gradually turns brown to dark brown and the pupal skin becomes harder.

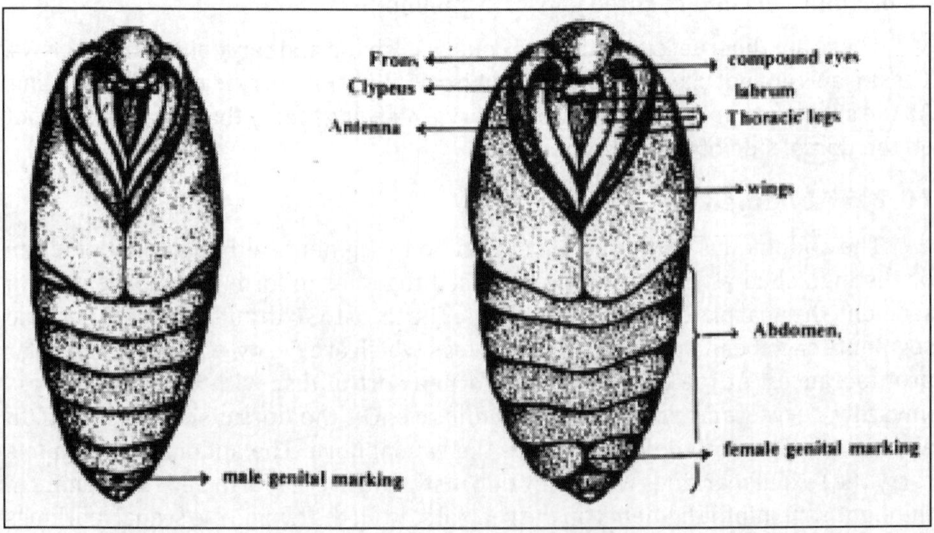

Male Female

Figure 1: Male and Female Pupa of *Bombyx mori*.

The prominent morphological parts visible are a pair of large compound eyes, a pair of large antennae, fore-and hind-wings and the legs. Ten of the abdominal segments can be seen on the ventral side, but only nine of them are visible on the dorsal side. Seven pairs of spiracles can also be discerned in the abdomen; the last pair is non-functional. The female has a fine longitudinal line on the 8th abdominal segment, whereas such a marking is absent in the case of the male.

10.2.8 Adult

The adult moth emerging from the pupa is incapable of flight as a result of its domestication for more than four thousand years. It does not feed during its short adult life. The body of the moth like that of the larva is composed of three distinct segments *viz.* head, thorax and abdomen.

Another important characteristic is common with other moths and butterflies is that the body surface is covered with scales. The compound eyes are situated on either side of the head. The ocelli are absent. The antennae are conspicuous, large and bipectinate.

The thorax consists of three segments namely pro, meso and metathorax as in the larva. The meso thorax is the largest and is the pentagonal. There are three pairs of thoracic legs, one pair on each of the three thoracic segments. Each of the thoracic legs is composed of five segments. The meso and metathorax bear two pairs of wings, the front pair overlapping the hind pair when the moth is in the resting position.

In the male eight abdominal segments are visible, in the female seven. There are six pairs spiracles present laterally on either side of the body.

The difference in the morphological characters in male and female adult silk moths

Male Silk Moth	***Female Silk Moth***
Smaller in body size	Larger in body size
Antennae larger in size, dark colour	Antennae smaller in size, pale colour
Abdomen is slender and small	Abdomen is stout
The caudal end has a pair of hooks known as harpes helping in copulation.	The caudal end has a median knob like projection with sensory hairs, which is protruded and retracted to expel the pheromones (Ovipositor)
Very active	Lethargic

Morphologically the female and male can be easily distinguished in the adult stage. The female has comparatively smaller antennae, its body and the abdomen are fatter and larger and it is generally less active than the male moth.

At the caudal end, the male moth has a pair of hooks known as harps whereas the female moth has a knob -like projection with sensory hairs. These differences help to a large extent in separating the sexes for preparation of hybrid eggs.

Male Moth **Female Moth**

Chapter 11

Silkworm Rearing Technology

11.1 Silkworm Rearing House

High profit is the ultimate aim of farmers to take up any cash crop. In sericulture, this can be achieved by producing good quality cocoons from 100dfls (Unit of silkworm seed or mulberry area). To meet the above demand, sound planning of the rearing house is necessary. Hoserver, the rearing house demands certain specified environmental conditions particularly temperature and humidity. Hence rearing house are planned to ensure optimum environmental conditions to get good quality cocoons.

The following factors are to be considered for annual planning

1. Crop wise leaf production per unit area
2. Rearing period
3. Number of crops per year
4. Rearing facilities
5. Labour management

Precaution to be taken for selection of rearing houses

1. Avoid dampness
2. Avoid stagnation of air
3. Avoid direct and too strong drought of air
4. Avoid exposure to bright sun light and radiation

5. Ensure an equable temperature and humidity
6. Ensure good ventilation

11.2 Chawki Silkworm Rearing

The young age silkworm rearing or chawki rearing is a vital aspect of sericulture industry. This supports the sericulture industry by harvesting the cocoon crop successfully. Chawki rearing is a must to reduce the crop loss and increase the cocoon yield. The advantages of co-operative chawki rearing of young silkworms are described below.

1. Stabilization of cocoon crop and increase in yield
2. Improvement in quality of cocoons
3. Control over incidence of disease
4. Reduction in rearing expenditure and
5. Utilisation of labour for other works

11.2.1 Plan for Model Chawki Rearing House (5000 dfls)

The model chawki rearing house should have one big hall and three separate rooms for the purpose of maintaining temperature and humidity. Each room can accommodate seven chawki rearing stand with sufficient working space. An average of 3500 dfls can be reared up to second moult in each room of 840 sq ft. The entire building plan should have two big rooms (18'x24') followed by leaf storage room (10'x11.3'), leaf chopping room (10'x12'), and store (7'x11.3'). In addition to all a separate office room (13.3'x12') and other amenities should be provided. In the plan, the total length (0-63') and width of 32.3' is recommended for 5000 dfls. The total ground plan of the building includes plinth (2032 sqft) and floor area of 1761 sq ft.

The young silkworm rearing room should be separate maintained near mulberry garden. Preferably with RCC roofing and provided with good ventilation. The building structure should be such that fluctuations in external environmental conditions should not affect the inside condition and thus the inner climate can be regulated. This can be ensured with thick walls or double wall with air pocket inside or with hollow bricks. Providing false ceiling and ventilators to make it convenient for rearing and maintaining the temperature and humidity.

11.2.1 Environmental Conditions

Environmental factors such as temperature, humidity, light and air current are influence on the growth of young silkworm. These should be maintained throughout the rearing period.

1. **Temperature**

   a. 20-28°C: This temperature range is harmless to the growth of silkworms. But higher or lower than this range is harmful to the silkworm.

b. 25°C: This temperature is favourable for the healthy growth.

c. 26-28°C: This temperature is favourable for making good quality cocoons.

In the early instars (I-III) should be relatively higher than that in the later instars (IV-V).

2. Humidity

High humidity causes reduction in the growing period of silkworm and acceleration in the physiological activities whereas; low humidity leads to prolongation of the larval period. A range of 85-90 per cent for 1st and 2nd instar and 80 per cent for third instar is ideal. The room is too moist, it becomes favourable for the pathogenic growth and cause diseases. During moulting humidity of 70 per cent is required.

3. Light

Silkworms are fond of dim light (15-30 lux) and avoid strong light and darkness. Light has little effect on silkworm health. 18 hrs photoperiod per day is found optimum for silkworms. In this condition the weight of cocoons and cocoon shell become heavier than those reared under dark condition.

4. Air Current

Fresh air is required for silkworms. In the rearing room, the air gets polluted due to accumulation of CO_2 released from the respiratory activities of workers and mulberry leaves, formaldehyde gas from disinfectants and ammonia from litter. The safe limit for these gases in silkworm rearing is CO_2 1-2 per cent, formaldehyde gas 1 per cent, SO_2 0.02 per cent and ammonia 0.1 per cent. A precautionary measure should be taken to avoid accumulation of above gases in chawki rearing is necessary.

11.3 Late-Age Silkworm Rearing

Rearing house for late-age silkworm rearing should have larger area with good ventilation for manipulation of humidity and temperature apart from provision for preventing uzi flies. The specification of rearing room depends on Method of silkworm rearing and rearing tools.

11.4 Plan for Rearing House

To rear late-age silkworms Rearers are constructing different types of rearing rooms according to the availabilities with them these rearing rooms are enable to maintain the optimum temperature humidity and air current for the growth and development of silkworm. However, CSR and TI Mysore recommended different models of late-age silkworm according to climatic conditions revealed in tropical, sub-tropical and cold reasons in the countries.

Basic considerations for construction of late-age rearing house

I. Selection of building site

II. Orientation of building

III. Size of the building

IV. Wall and roof structure of building

I. Selection of Building Site

1. The rearing site should not be water logged.
2. In high rainfall area a dry, sunny, well ventilated and well drained land should be selected.
3. Avoid too stagnant air due to thick vegetation all around.
4. Avoid too much open area for strong winds

II. Orientation of Building

In tropical regions the orientation of rearing house is very important as it affects the temperature and humidity of room. The best orientation of building, doors and windows is north-south walls. The width of the building is east-west direction.

III. Size of the Building

The size of building depends largely on quantum and type of rearing (shelf/ stand rearing and floor rearing). In floor area 2 sqft/dfl is required for tray rearing and 4sq ft/dfl for shoot feeding method.

IV. Wall and Roof Structure of Building

The wall and roof structure of the building depending on the availability of raw materials and environmental conditions. The materials required for construction of wall *viz.,* burnt brick, kacha brick, and mud or spilt bamboo pieces plastered with on either sides. The materials used for roof construction are country tail, Mangalore tile and thatch.

The area requirement for (200-300 dfls) shoot rearing house consist of

1. Mulberry storage room

   10'x12' = 120 sqft
2. Rearing room

   40'x18' = 720 sqft
3. Mounting room

   10'x12' = 120 sqft

   Total area = 960 sqft

11.5 Study of Rearing Equipments/Appliances

Following are the appliance required for rearing silkworms for acre of mulberry garden (300dfls)

11.5.1 Equipments (Non-recurring (general)

1. **Disinfection mask and protective gum shoes (1 set)**

These are used during spraying of formalin to protect the eyes and skin of the rearer from corrosive effect of formalin.

2. **Sprayer (power sprayer, multipurpose pump or rocking sprayer for disinfection)-1No.**

These are used for spraying formalin/sanitech during disinfection depend upon the quantity of spraying solution (concentration, floor area and type of disinfection).

3. **Room heater with thermostat (2KV) or charcoal stove (3Nos.)**

These are necessary for maintaining optimum environmental conditions required for silkworm in the rearing house depending upon the weather variation these are employed in different rearing season.

4. **Water air cooler (heavy duty) -3Nos.**

This is very commonly used in silkworm rearing house to maintain the optimum relative humidity required for silkworms during summer season.

5. **Kerosene blow lamp- 1No.**

This is used in rearing house when there is no power to carry out different rearing operations.

6. **Wet and dry bulb thermometer- 3Nos.**

This is very much required to measure temperature and relative humidity of rearing house and also in the rearing bed.

7. **Forceps 6" long- 3Nos.**

This is necessary to lift worm rearing house and increased rearing bed house avoid contamination from house.

8. **Plastic sieve (1No.)**

This is used in rearing house for uniform dusting of lime powder on the silkworm body.

115.2 Non-recurring (specific)

1. **Rearing Building**

A separate rearing building with specific dimension inclusive of leaf preservation cum preparation room with ante room is suggested.

2. Egg Transportation bag-1No.

Silkworm eggs are distributed to the farmers for commercial rearing when active development of embryo is in process. This silkworm eggs transported from grainage during coolers hours of the day in an egg transportation bag. The bag is specially designed and black in colour to maintain high humidity and low temperature by wetting the surface.

3. Egg incubation chamber-1No.

In India it is designed with an auto control of temperature and relative humidity in chamber which is double walled metallic box with insulating materials between the walls provided exhaust fan and heater with thermostat control.

4. Egg incubation chamber-1No.

During incubation of loose eggs, these frames are used. Loose eggs are spread in a single layer and covered with tissue paper till the day of hatching after ensuring optimum temperature and humidity. Generally they are made out of plastic.

5. Black box-1 No.

This is commonly employed after body pigmentation stage to inhibit the growth of embryo to certain extent and to accelerate the embryonic growth which is not in order. (Lag in development) preferably 48hrs before hatching is advisable.

6. Chawki rearing tray (wooden/plastic, 3'x4'x3")-24Nos.

To keep I-III instar silkworms.

7. Chawki rearing bottom stand-2Nos.

To rest all chawki rearing trays (10-12) for each stand during rearing.

8. Feeding stand -2Nos.

During feeding the trays are placed on feeding stand for convince and uniform distribution of leaves in the bed.

9. Ant well-72Nos.

To prevent ants.

10. Leaf chopping board-1No.

It is necessary to give optimum size leaf bits for chawki worms to cut the leaves chopping board is employed.

11. Leaf chopping knife-1No.

Used for cutting and trimming of shoots before feeding.

12. Leaf mat (4'x6')-2Nos.

After leaf is harvested it is necessary to store without dust and dirt in the rearing house by keeping the leaves on leaf mat. To avoid withering of leaves.

13. Leaf chamber for chawki-1No.

Leaves are preserved in the chawki room without affecting the quality of the leaf in the chamber (clean germ free area under high humidity and low temperature). This is usually prepared by wood covered with gunny cloth/polythene sheet to maintain freshness for longer time frequent sprinkling with water is advisable.

14. Bed cleaning net (3'x4', mesh size ¼ inch x ¼ inch)-48Nos.

Different size nets are being employed to separate worms from rearing bed to ensure effective bed cleaning operation.

15. Earthen pot-1No.

To preserve the leaves and eggs during incubation.

16. Litter basket-2Nos.

To carry wastage without contaminating room for easy disposal.

17. Late age rearing tray (bamboo 3.5' diameter for leaf feeding method only)-2 Nos.

For rearing late-age silkworms.

18. Rearing stand-12 for CB and 14 for BV

To rest rearing trays in 10-12 tiers on the rearing.Theya re made out of iron, steel and wood.

19. Feeding stands (3)

They are made out of wood, steel or iron. Used to feed the worms which are in to top (last tier)

20. Shoot rearing rack 2 (5'x35'x3')

These are used for shoot feeding method, usually (3-4=1 ½', 1' and 1') are employed depend upon number of dfls brushed.

21. Moutages (Chandrike) 120

These are made out of bamboo, which are used for spinning the cocoon.

22. Plastic basin (big, 18"d) Plastic buckets with lid (small and big 10 and 20lit capacity-2Nos.)

These are required for disinfect the rearing houses after completion of one crop.

11.6 Recurring

Requirement per Crop

1. Paraffin paper (20m/crop)-20m
2. Old news paper -10kg
3. Formalin-7kg

4. Bleaching powder- 7kg
5. Lime -25kg
6. Bed disinfectant-12kg
7. Power and fuel

11.7 Disinfection and Hygiene

Disinfection is the act of destroying disease causing germs, Disinfectant is an agent that frees an organism form infection/Disinfection forms an integral part of healthy and successful rearing. To ensure total destruction of disease causing microorganisms *viz.*, Bacteria, fungi, viruses and protozoa. Disinfection of rearing room and appliance is the first step towards silkworm rearing. Inactivation or removal of disease germs and to prevent silkworms from diseases. The word "disinfection" refers to the use of germicidal chemical agents to destroy the potential activity of an organism.

Disinfection before and after rearing involves the rearing room and equipment and can be done thoroughly by using stronger disinfectants. Disinfection during silkworm rearing has to be done carefully without affecting the healthiness of silkworm.

11.7.1 Disinfection Methods

Physical Disinfection

- ✰ Burning
- ✰ Burying
- ✰ Exposing to sun light
- ✰ Exposing to dry heat/steam
- ✰ Exposing to UV radiation

Chemical Disinfection

Exposing to different Chemical disinfectants

- ✰ Formalin
- ✰ Bleaching Powder
- ✰ Chlorine dioxide
- ✰ Slaked lime
- ✰ Lodophores

11.7.2 Common Disinfectants Used in Sericulture

I. Formalin (Formal)

- ✰ 36-40 per cent solution of formaldehyde in water

- ☆ 10-15 per cent methanol added to prevent polymerization
- ☆ Colourless liquid with pungent odour
- ☆ A powerful reducing agent
- ☆ Takes oxygen from germ cells and gets itself oxidized to Formic acid
- ☆ Action faster at temp. above 25° C
- ☆ Used in 2 per cent concentration
- ☆ Near air tight condition required for best result

II. Bleaching Powder (Calcium Oxychloride-Ca (OC1)2)

- ☆ Otherwise known as chlorinated lime or chloride of lime
- ☆ Dull white amorphous powder
- ☆ Hygroscopic – absorbs moisture on exposure to air and rapidly decomposes
- ☆ Relatively unstable chlorine carrier in solid form. Maximum available chlorine approaches 39 per cent
- ☆ Commercial products have between 24-37 per cent available chlorine.
- ☆ Action in Three ways
- ☆ Oxidation-releases nascent oxygen
- ☆ Releases chlorine
- ☆ Calcium hydroxide formed in the process is germicidal.

III. Chlorine Dioxide (ClO_2)

- ☆ Strong oxidizing agent
- ☆ Unpleasant odour of Chlorine
- ☆ Available in 20000 ppm concentration as a commercial product, Sanitech
- ☆ Less corrosive, compared to bleaching powder
- ☆ Highly germicidal

IV. Slaked Lime (Calcium Hydroxide – $Ca(OH)_2$)

- ☆ Dull white powder
- ☆ Prepared from burnt rock/shell lime by sprinkling water and subsequent pulverizing
- ☆ Good anti viral activity

V. Paraformaldehyde (Tri-oxymethylene)

- ☆ White granular substance
- ☆ Releases formaldehyde vapour

- ☆ Formed by polymeraisation of formaldehyde
- ☆ Effective for fumigation
- ☆ On heating, sublimates, releasing formaldehyde gas
- ☆ One of the major ingredients in certain silkworm bed disinfectants

11.7.3 Preparation of Disinfectant Solutions

2 per cent Formalin (for closed type of rearing house)

Prepared from Commercial Formalin according to the following formula.

$$\frac{\text{Available conc. of formaldehyde} - \text{Required conc. of formaldehyde}}{\text{Required conc. of formaldehyde}}$$

= Parts of water to be added to one part of formalin

Bleaching Powder (for open type of rearing house)

- ☆ 2 per cent solution of high grade bleaching powder is prepared
- ☆ Mix 20g of bleaching powder in 1 litreof water
- ☆ The mixture is filtered and the clear solution is used for disinfection
- ☆ 0.3 per cent slaked lime is added (@ 3g/litre) to it to improve the disinfection ability

Chlorine Dioxide (Commercially available as Sanitech)

- ☆ Chlorine dioxide is available as Sanitech' in 20,000 ppm concentration. 500 ppm solution is needed for disinfection purpose.
- ☆ 50g of activator crystals is added to 500 ml of Sanitech and dissolved till the colour of the solution changes to yellow (Solution A).
- ☆ Mix 100g of slaked lime powder in 19.5 liters of water (Solution B).
- ☆ Mix Solutions A and B, *i.e.* 500 ml + 19.5 liters = 20 liters

5 per cent bleaching powder + slaked lime mixture

- ☆ High grade bleaching powder is mixed with fine slaked lime powder of 200-250 mesh size @ 50g per 950g of lime (1:19).
- ☆ The mixture is used for dusting at the entrance and around the grainage building to maintain proper hygiene.

Quantity Requirement of Disinfectants

Quantity is estimated on the basis of surface area of equipments and buildings. In the rooms, the surface area of roof, floor and four sides are taken into account.

Area of floor and roof	LxBx2
Area of 2 side walls	LxBx2
Area of other 2 side walls	LxBx2

Total Area (Sq.mtrs)

The disinfectant solution should be prepared at the rate of l litre/2.5 sq. meters (or 2 litre/sq.meters of floor area).

Disinfection Procedures

Item	*Disinfectant*	*Quantity*	*Method*
Grainage building/ Rearing House	2 per cent formalin + 0.05 per cent detergent	1 lit./2.5 sq.mt. area OR 1 lit./2 sq.mt floor area	Drenching Spray
	2 per cent bleaching powder solution + 0.3 per cent slaked lime	-do-	-do-
	500 ppm Chlorine dioxide	-do-	-do-
Grainage/ Rearing Equipments	2 per cent formalin + 0.05 per cent detergent	1 lit./2.5 Sq.mt. area OR 1 lit./2	Drenching Spray-Equipments are either dipped in a disinfection tank or piled up, sprayed with disinfectants and covered with vinyl sheet for 30 min.
	2 per cent bleaching powder solution + 0.3 per cent slaked lime	-do-	
	500 ppm Chlorine dioxide	-do-	
Building Surroundings	5 per cent bleaching powder in slaked lime mixture	200 g/Sq.mt.	Sprinkled at the entrance and surrou-nding areas of the building on alternate days.
Moth testing table and platform	2 per cent bleaching powder solution	As required	Wiping before and after moth testing work.
Silkworm eggs	per cent formalin for sheet eggs and 0.3 per cent bleaching powder for loose eggs	-	Egg sheets are dipped for 10 min and dried in shade

11.8 Hygiene Measures

11.8.1 Rearing/Grainage Hygiene

- ☆ Disposal of rearing/grainage wastes into a soak pit
- ☆ Burning of wastes in incinerator
- ☆ Mopping the rearing/grainage room floor with 2 per cent bleaching powder solution after every day's operation.
- ☆ Application of bed disinfectant in rearing
- ☆ Disposal of infected larvae as and when noticed

11.8.2 Personal Hygiene

- ☆ Disinfection of hand and feet of grainage/rearing workers before and after the daily operations- by washing with carbolic soap/detol.
- ☆ Wearing of aprons while carrying out different operations.

11.9 Quality of Disinfectants

Quality of disinfectants is a very important aspect determining the effectiveness of disinfection.

In the case of formalin it is the strength of formaldehyde present in the commercial formalin while in bleaching powder it is the percentage of available chloring that determines the quality.

The quality of these disinfectants can be assessed by simple chemical analytical methods.

11.10 Step-wise Approach for Proper Disinfection

- ☆ Grainage appliances are kept outside in bright sunlight for a period of 5-6 hrs.
- ☆ Appliances are disinfected by dipping them into 2 per cent bleaching powder solution for 30 minutes followed by sun drying. Alternately, the appliances can be given a drenching spray with 2 per cent formalin or 2 per cent bleaching powder solution and kept completely covered with a vinyl sheet for at least 30 min.
- ☆ The rooms in a grainage are cleaned with 2 per cent bleaching powder solution (by spray).
- ☆ The rooms are later disinfected by spraying 2 per cent formalin along with 0.05 per cent detergent and kept closed for 24 hrs.
- ☆ The surrounding areas of the grainage building are disinfected by sprinkling 10 per cent bleaching powder + slaked lime mixture at the rate of 200g/sq. meter, 1-2 meter area around, on alternate days.

☆ Disinfection may be followed by fumigation, if the buildings could be made completely air-tight. Formalin (36 per cent) is diluted in the ratio of 1:1 with water and heated over a small flame, Fumigation is effective only under high humid conditions.

Points to Remember

☆ Integration of chemical and physical methods combined with strict hygiene measures is essential for achieving the objective of complete extermination of disease causing germs from the environment.

☆ Usage of different types of disinfectants alternately will be helpful in combating the problem of microbes developing resistance to a particular chemical due to continuous exposure.

Disinfection and Hygiene in Silkworm Rearing

	Reshamkeet Oushadh	***Sanjeevini***	***Suraksha***	***Resham Jyothi***	***Vijetha***
Recom-mended for prevention/ control of	Muscardine and Grasserie	Grasserie and Flacherie	Muscardine	All diseases	All diseases
Inventor	CSRTI, Mysore	KSSRDI, Bangalore	KSSRDI, Bangalore	SSTL, Bangalore	CSRTI, Mysore
Cost/kg (Rs.)	15/-	20/-	20/-	35/-	35/-
Shelf life	6 months	6 months	6 months	1 year	1 year
Quantity required	5-6 kg/100 dfl's	5-6 kg/100 dfl's	5-6 kg/100 dfl's	5-6 kg/100 dfl's	5-6 kg/100 dfl's
Mode of application	Dust on silkworm body and roaring seat once after every moult and on 4th day of final instar after bed cleaning	Thinly Dust on hatched larvae and on empty tray while brushing. Dust on silkworm body and rearing seat once after every moult and on 4th day of final instar after bed cleaning.	Thinly Dust on hatched larvae and on empty tray while brushing. Dust on silkworm body and rearing seat once after every moult and on 4th day of final instar after bed cleaning.	Dust on hatched larvae. Dust on silkworm body and rearing seat once after every moult and on 4th day of final instar after bed cleaning.	Dusting once after every moult and on 4th day of final instar after bed cleaning.

11.11 Egg Incubation Methods and Black Boxing

Incubation is a process in which the silkworm eggs preserved at ideal conditions to avoid irregular hatching. Incubation is also known as "seed warming" by which the activated silk eggs are protected in proper temperature and environmental conditions so that the embryos can develop normally and the eggs hatch uniformity. The ideal incubation is to maintain a temperature of 25°c and 75 per cent to 80 per cent humidity continuously. The eggs are kept spread in wooden trays provided with paraffin paper for base and the cover with wet foam pads all around inside the trays.

11.11.1 Different Methods of Egg Incubation

a) Incubation Chamber

This is made out of an iron sheet with a double layered wall in which water can be poured to maintain the humidity. It has a ventilator on top and an electric heater or coal furnace at the bottom to maintain the optimum temperature.

b) Incubation Room

The incubation room is used when a large number of eggs are incubated. The incubation room must be built in such way to maintain uniform temperature and humidity and it should have sufficient illumination and good ventilation. To maintain the required temperature and humidity conditions, silkworm eggs are incubated in BOD incubator or spread in a wooden tray in which wet foam pads are placed at the periphery and covered with paraffin paper.

c) Earthen Pot

Well baked earthen pot with a wide mouth of about 15" diameter is chosen. The pot is filled with a clean sand layer of 2-3" thick and made wet depending on the need. It is provided with a row of holes of about half an inch diameter and situated 2-3" above the sand level. It is covered with a clean wet cloth. The sand is kept wet one day prior to incubation. The layings are tied to a rod and inside the pot and the pot is covered with a wet cloth and placed over an ant well. Monitor the cloth/sand regularly for wetness. As a result the temperature is reduced by 5-10°C.

d) Double Brick Walled Chamber

The chamber is constructed using burnt bricks, sand and cement. There is gap of about 3" between inner and outer walls and is filled with loose clean sand. While constructing a wire mesh shutter may be fixed at the top to avoid the entry of rodent and other predators. The standard size of the outer wall is 6'x4'x3'. The inner wall is 4'x2'x3' with a gap of 3" between them. This incubation device accommodates 5000-6000 dfls. Egg sheets are tied to a metal rod with sufficient gap between them and are hung in the chamber. The sand layer has to be soaked periodically with clean water depending upon the environmental conditions. The temperature and humidity conditions inside the chamber are on par with that of earthen pot.

11.11.2 Black Boxing

The black boxing is done to achieve uniform hatching on the same day. It is done on two days or one day before hatching when eye spot or blue egg stage is noticed.

- ☆ Remove the paraffin paper covered to the tray with incubated DFLs.
- ☆ Cover the entire tray with black cloth and keep undisturbed.
- ☆ Remove the black cloth and keep the layings near the light source for hatching

OR

- ☆ Wrap the egg sheets in a tissue paper one day prior to hatching
- ☆ Transfer such DFLs into a black paper cover or black boxes
- ☆ Place the covers or boxes under required humidity and temperature conditions.

Brushing of Silkworms and Selection of Leaf for different Instars

Transfer of the newly hatched larvae from egg card/egg case on to the silkworm rearing bed is called as brushing. Rearing of young age silkworms up to II moult is called chawki rearing which usually lasts up to 9 to 10days. Rearing of young age silkworms under ideal environmental conditions, feeding succulent, nutritious, tender leaves makes the larvae grow robust and make more stress tolerant during late age rearing.

a) To acquaint with brushing techniques
b) Tray method, earthen pot method, Double walled chamber method and Low cost incubation chamber methods.

Selection of Leaf for Different Instars

Mulberry leaves are the natural food of the silkworm, *B.mori* L. but for the purpose of feeding silkworms the plants have to be specially grown to ensure a harvest of leaves that specially meet the nutritional and feeding requirements of the different instars of silkworms. The nutritive value of mulberry leaf decreases with position of leaf from top to bottom of the stem.

Infant worms require soft leaves rich in water content, proteins, carbohydrates and minerals, while mature worms need mature leaves with less water and sugar content but with more protein and crude fibresi, they required less mature worms. If young worms are given mature leaves, unequal growth results because their mandibles cannot cut the coarse leaves and their digestive enzymes are too weak to digest the higher starch contents. If mature worms are given tender leaves, due to inadequate nutrition they become susceptible to diseases. Mature leaves are given to late-age worms

Optimum Leaf for Feeding Young-age Silkworm in different Seasons

Season	*Plant Growth*	*Larval Instar*	*Leaves Selected*
Summer and early autumn	Vigorous	Instar I	4-5 leaves below the largest glossy leaf including the largest glossy leaf.
		Instar II	6-7 leaves below the largest glossy leaf including the largest glossy leaf.
		Instar III	7-8 leaves below the largest glossy leaf, excluding one or two leaves just below the largest glossy leaf.
Late autumn	Poor	Instar I	3-4 leaves below the maximum light-exposed leaf including the largest glossy leaf.
		Instar II	4-5 leaves below the maximum light-exposed leaf including the largest glossy leaf.
		Instar III	5-6 leaves below the maximum light-exposed leaf and excluding it.

For the first instar, leaves lying between the leaf at yellow lenticles and the leaf just above the axillary bud with a brown tip are selected.

For the second instar, all the leaves lying in-between the leaf at brown lenticles and the leaf at the axillary bud with a brown tip are selected. If the mulberry is not sufficiently exposed to sunlight, the buds often have a longitudinal line running towards the lower direction from the axillary bud with brown tip. In this case, all the leaves may be plucked.

For the third instar, leaves below the level of those selected for instar I are plucked, though for the third instar, shootlet feeding is preferred to individual leaf feeding.

The ideal leaves for lateage worms are those at the bottom of the stem. These are mature, thick, and slow to wilt rich in protein and have comparatively low moisture content.

11.12 Bed Cleaning, Moulting Worms and Management

This rearing operation is indispensable for the maintenance healthy and hygienic microclimate in the rearing bed. For ideal hygienic rearing, the bed should be cleaned before each feeding. The following schedule of bed cleaning is observed in India.

Instars	*Time*	*At what time*
Ist	Once	The third day after brushing when the larvae settle for the first moult.
IInd	Twice	1. Just after the first moult at the time of the first feeding.2. Before the larvae enter the second moult.
IIIrd	Thrice	1. After the second moult.2. In the middle of the instar.3. Just before the third moult.
IVth and Vth	Once	2nd day or Middle

IVth and Vth instar for shelf rearing, once a day and for floor rearing, once a day and for floor rearing, once for each instar.

11.12.1 Methods of Bed Cleaning

1. Rope Method

Depending upon the length of the shoot rearing rack, 2 or 3 ropes of 6 feet length are spread parallel to one another on the rearing bed leaving a margin of 1.5 feet on both sides. After two feeding (by providing the shoots across the ropes), most of the larvae climb and reach the surface layer of shoots, both the ends of the ropes are lifted together and transferred. This helps in separating the old bed easily without causing damage to the larvae. After removing the lod bed, the bundle of fresh shoot with larvae are opened and spread. The ropes remain below the bed which helps in removing the bed when all the larvae are mounted for spinning.

2. Net Method

In this method, nets of 5'x6' size with 2"x2" mesh size are spread over the rearing bed and after 2 feedings, most the larvae climb and reach the surface layer of shoots, nets are lifted and bed cleaning is done. The nets are spread on the racks and larvae are distributed as per the required area. Nets are left below the bed which will help in removing the bed while doing final cleaning. After cleaning the bed, disinfectants should be dusted over the rearing bed to check the spread of contamination.

11.12.2 Moulting Worms and Management

The silkworm *Bombyx mori* undergoes four moults and has five larval instars. The care taken during moulting time helps to insure inform growth of the silkworms. Under optimum rearing conditions the worms take 3 to 3½ days to settle for I and III moult and 2 to 2½days for II moult and takes 4 to 4½ days for IV moult. Moulting duration is 18 to 24houts.

Characteristics before Moult

1. The leaf consumption by silkworm reduces and it appears to be inactive.
2. Feed behavior.

3. Size of head appears to be smaller in relation to the size of the body.
4. Skin will be translucent, tight, smooth and shinny worm may spin small quantity of silk.

11.13 Study of Chawki and Late-Age Silkworms Rearing

The fourth and fifth instar of silkworms is known as late age worms and their rearing management differs from that of young age worms. During this period, silkworm body weight increases by 20 times, body volume by 29 times and silk gland weight by 200times. The ecological and nutritional requirement of late age worms require cooler and drier climate. They are voracious feeder and consume 90 per cent of total leaf consumption in this stage compared to V instar, the resistance of fourth instar to viral disease is very poor and these worms donot readily adapt to temperature lower than 20°C. Silkworms have lost the capacity to adjust to fluctuations in environmental conditions due to domestication and indoor rearing.

11.14 Different Methods of Chawki and Late-Age Silkworm Rearing

The young age silkworm rearing is a vital aspect of sericulture industry for the development of healthy and harvesting the cocoon crop successfully. In developed countries like Japan, China and Korea about 95 per cent of the farmers receive chawki worms but in India it is only 10-15 per cent. Therefore, it is desirable to have a separate rearing house/room for young silkworms preferably with RCC roofing provided with good ventilation. It would be convenient to provide false ceiling and ventilations required for maintenance of silkworms.

Different Types of Chawki Rearing

1. Paraffin Paper Method of Rearing

The young worms are placed on a sheet of paraffin paper and covered with another sheet of paper. The rearing bed covered with wet foam rubber strips to increase humidity. During moulting the paper is lifted to receive fresh air. The same may be repeated for feeding operation.

2. Box Rearing

This method is practised when temperature and humidity are lower. Wooden or plastic trays of 4'x3'x4" or 3'x2'x4" are used and arranged vertically. About 60g of eggs, up to the first instar or 20g up to the second instar can be reared in each box. The trays are arranged in cris-cross pattern for allowing the rearing bed to dry up.

3. Stand Rearing

This method is very commonly employed method when, optimum temperature and humidity are available or temperature is high in the rearing house. Young-age rearing is done in trays as it is practiced for late age rearing. Paraffin paper is used as seat and cover for young silkworms. This is not suitable for large scale rearing because it requires more space.

4. Cellar Rearing

In this Method the compact wooden cells are fixed with stands having 12 tiers for keeping rearing trays. A wooden door in the front is closed after the trays are inserted in cells. Warmth is maintained in the room either by electricity or a charcoal stove, and wet mats are spread on the floor for retaining humidity of all the three methods.

5. Isolation Chamber Rearing

The isolation chamber rearing is useful when temperature and relative humidity are very low. This chamber is made out of plywood sheets with wooden frame of size 6'x4'x5' with double door provided with heater and humidifier inside. It contains thermostat and humidistat to maintain optimum temperature and humidity. Rearing is done inside the chamber by filling up the trays on a chawki stand or on a small rearing stand. The paraffin paper is used for covering the rearing bed should be new and of good quality to withstand the melting point (50°C). Chamber provides with ventilators on the upper and lower side. This accommodates 300 dfls till second moult.

In addition to these box with lid, net method, low cost environator, single feeding method are very common methods of India. Whereas, Plastic tent rearing, Damp proof paper rearing, Hotbed rearing, Electric hotbed rearing are common in China, Russia and Ukraine but, in Japan Pack rearing, Wet cloth rearing, Single feed rearing, Kasso rearing, Showa rearing are practiced by individual farmers. In addition to above, the co-operative rearing like Tenryu method, Gumma method, Electrically heated floor, Nagano style, Saitama style, Mechanical rearing and Rearing on artificial diet are common methods established by Japanese.

Late-Age Silkworm Rearing

Rearing of IVth and Vth instar is considered as late-age silkworm rearing or grown up silkworms. It is an easy task as compared to young-age rearing. They need more labour and attention to harvest good quality cocoons, failing which all the activities get altered and leading to poor yield. During this stage the silkworm body volume increased by 29 times, body weight by 25 times and silk gland weight by 200 times. The requirements of ecological and nutritional conditions are different from young age.

Characteristics of Late-Age Silkworm

1. They have low resistance tolerance to high temperature, high humidity and poor ventilation.
2. The amount of mulberry ingested and digested greatly increases. They can consume coarser and wilted leaves, but the ingestion-digestion ratio is lower than that of young age silkworms.

3. They can thrive/withstand on reduced amount of mulberry leaves that will affect the quality of cocoons to a considerable degree.
4. They are relatively more resistance to chemicals.

Methods of Late-Age Rearing

The rearing of late-age worms (IV and V instar) is conducted by various methods and farmers adopt the methodology suitable to the space available for rearing infrastructure and economic condition. Thus the following types are very common.

a. Tray rearing

b. Shoot rearing

c. Floor rearing

a. Tray Rearing

In this method, leaves are plucked from the bushes/trees and then fed to silkworm whole or cut as per the growth of silkworm. The rearing is done in wooden/bamboo trays in 10-12 tier stands. It is more laborious. The tray rearing method is very popular as it needs less space.

b. Floor Rearing

This method is adopted for late-age rearing by leaf feeding or shoots feeding. If farmers have enough space for rearing and can maintain hygienic conditions, rearing on floor can be conducted. Two or three beds can be prepared with 1 to 1.5m width and length depending upon the space available *i.e.*, 5-7m in the farmers rearing house, leaving about 1-1.5m space between the beds and from the margin to attend the feeding work.

c. Shoot Rearing

In this type of rearing, the whole shoots of mulberry are used after III moult. It is also evident that, this method needs less labour. As there are three dimensional distributions of worms, it creates better aeration in the bed besides accommodating 50 per cent more silkworms per unit area of rearing bed as compared to the leaf rearing. In tropical countries, 3 tier racks is advisable. The bed cleaning is practiced once each in fourth and fifth instar.

Study of Shoot Feeding Method and Self Rearing Method

Shoot feeding method is introduced for last two stages of rearing by giving whole mulberry shoots instead of individual leaves, and it is carried out on platform type rearing rack. There will be one bed cleaning after fourth moult. In shoot rearing, feeding is done 2 or 3 times per day according to the season. It is observed that most of the farmers are switching over from stand rearing to shoot rearing due to higher labour efficiency.

Advantages of Shoot Feeding

1. Labour saving up to 50 per cent compared to leaf feeding method.
2. Leaf saving up to 15 per cent.
3. Cocoon character and ERR are better.
4. Better keeping quality of mulberry leaf during storing and in the rearing bed.
5. Better hygienic conditions can be maintained.
6. Handling of the silkworm is minimised, hence contamination and spreading of disease reduced.
7. Bed cleaning is required only after IV moult.
8. Worms and leaves do not come in contact with the litter, hence chances of secondary contamination is minimised.
9. Better aeration in the bed is ensured.
10. Cost of production is reduced.

Disadvantages of Shoot Feeding

1. Requires more floor area (about 30 per cent) as compared to tray rearing.
2. Bed refuses will not be available as cattle feed.
3. Planting material will not be available for new plantation.

Cleaning

Shoot rearing is an easy and labour saving method since frequency of feeding, and bed cleaning is considerably reduced. One time cleaning on 2nd day of Vth instar is advised. In shoot feeding bed cleaning can be done by two methods.

1. Rope Method

Depending upon the length of the shoot rearing rack, 2 or 3 ropes of 6 feet length are spread parallel to one another on the rearing bed leaving a margin of 1.5 feet on both sides. After two feeding (by providing the shoots across the ropes), most of the larvae climb and reach the surface layer of shoots, both the ends of the ropes are lifted together and transferred. This helps in separating the old bed easily without causing damage to the larvae. After removing the old bed, the bundle of fresh shoot with larvae are opened and spread. The ropes remain below the bed which helps in removing the bed when all the larvae are mounted for spinning.

2. Net Method

In this method, nets of 5'x6' size with 2"x2" mesh size are spread over the rearing bed and after 2 feedings, most of the larvae climb and reach the surface layer of shoots, nets are lifted and bed cleaning is done. The nets are spread on the racks and larvae are distributed as per the required area. Nets are left below the bed which will help in removing the bed while doing final cleaning. After cleaning the bed, disinfectants should be dusted over the rearing bed to check the spread of contamination.

Self Rearing Method

In this method of rearing, the worms reared are kept in trays stacked on a shelf. The main advantage of this method is to accommodate more number of worms per unit area. But the only disadvantage is that, it requires more time and labour for feeding, bed cleaning and other rearing operations. It is most popular method of rearing in all countries where cheap labour is available.

Rearing Stand

In this method, the silkworms are reared in circular or rectangular trays stacked one above the other in vertical rows on a rearing shelf or stand. The material used and the size of the stand vary from country to country. In general, locally available wood or bamboo is used for making the stand as well as the trays. A typical stand has a height of 2.5m, length 1.5m and width 0.65m, with 10-12 cross bars at a distance of 0.15m to accommodate 10-12 tiers or rearing trays.

Rearing Trays

Circular or rectangular trays made of cheap but long-lasting locally available material amenable to disinfection is used. Mostly wood or bamboo is used. Trays for young-age worms are solidly constructed so that moisture and heat are retained within while those for late-age worms are loosely constructed to allow more aeration and prevent carbon dioxide from accumulating in the rearing bed. Circular trays are 138 wm in diameter and about 6.5 cm deep. Rectangular trays have a standard size of 1.2m x 0.9 used temporarily. Wood or bamboo is replaced by cheap plastic or other materials that are not subject to bacterial attack as is wood.

Silkworm Ripening

The final instar/V instar larva stops feeding after attaining maximum growth and ready to spin the cocoon is called as ripe worm/mature worm. This stage is obtained after 6-7 days and 7-9days of feeding in fifth instar for MV and BV races, respectively. Transferring of mature silkworm to a suitable frame/mountage to spin cocoons is called mounting.

Identification of Ripen Worms

1. The appetite reduces and uneaten leaves are found at feeding.
2. The faeces become softer and can be crushed with fingertips.
3. In the newly mature silkworms, the front part of the body looks little transparent.
4. Worms move their anterior half of the body to front and back keeping the caudal portion of the body supported firmly with abdominal legs, searching for a place to spin cocoons.

5. Larvae start moving to the periphery of the rearing beds.
6. Secrete small quantity of silk to form hammock.

Methods of Mounting

1. Hand Picking

In this method mature larvae are individually picked one by one by hand and transferred on to mountage.

2. Simultaneous Mounting (Free mounting)

This is similar to hand picking, when 20 per cent of larvae are found mature are mounted on the mountage. The difference is that, instead of mounting only the ripe worms, all the worms in the bed are simultaneously collected and mounted.

3. Net Method

When ripe worms are noticed straw, rope, mat, brush type of mountages and cleaning nets are spread above the rearing beds. The only ripe worms crawl on to the mountages are mounted.

4. Frame Mounting

When a large number of larvae are found mature, a branch of mulberry leaf is supplied over the frame the mature larvae crawl (rise above the frame) on to the mulberry leaves are collected and transferred to the cocooning frame. The left out worms were handpicked and transferred on to the same frame.

5. Natural Mounting (Shelf Mounting)

This method is slightly advanced as compared to the net and frame methods of picking. When mature larvae appear initially they are picked up one by one and transferred to the mountage. Later, when half of the batch found to be in a mature state, mulberry is supplied and the cocooning mounting gently kept over the rearing bed. This will be kept for the 250 works collect on it. The mountage is than removed and positioned as a stand in the rearing room. This will be repeated several times remaining ones hand picking.

6. Jobarai Method (Shaking of shoots)

In this method mature worms are collected by shoot shaking, the worms which are shaken on a vinyl sheet manually on stand. The worms on mulberry shoots are collected by vibrating machine on a jobarai table. The worms that are collected are mixed with faces and bed refuses are cleaned by using net.

11.15 Climatic Condition Required for Silkworm Rearing

Silkworms have lost the capacity to adjust to fluctuations in environmental conditions due to domestication and indoor rearing. The important factors to be regulated include temperature, humidity, light and airflow.

Temperature

Larval Stage	*Temperature*
Ist Instar	26-28°C
IIst Instar	26-28°C
IIIst Instar	24-26°C
IVst Instar	24-25°C
Vst Instar	23-24°C

Regarding rearing temperature of each instar, it is advisable that the temperature in the early instars (I-III) should be relatively higher than those later instars (IV-V). In case sufficient amount of nutritive mulberry leaves are not given to silkworms, high temperature will be harmful to the physiology of silkworms there by injuring the health leading to poor quality cocoons. Lower the temperature affects the metabolic process in silkworm and higher the temperature reduces the cocoon production. To adopt to remove extra heat from within the room by water evaporation is to hang wet gunny bags or straw rush mats on the door and window which also increase the humidity within room. These can also serve the same purpose.

Humidity

High humidity shortens larval duration by accelerating the physiological functions of silkworms and low humidity lengthens the growing period of silkworms, especially young-age silkworms.

Larval Stage	*Humidity*
Ist Instar	90 per cent
IIst Instar	85 per cent
IIIst Instar	75 per cent
IVst Instar	60 per cent
Vst Instar	70 per cent

In general, pure lines and multivoltines require lower temperatures than bivoltines at the same stage. The optimum humidity should also be adjusted to temperature conditions.

Electrically operated humidifiers may be used for maintaining desired humidity. To lower humidity, lime is sprinkled on the floor and to increase air circulations, windows and doors are opened and to increase humidty, windows and doors are closed, wet gunny bags are hang, and water is sprinkled on the leaves.

Ventilation

Silkworm breathes on oxygen from the atmosphere in the room and gives off carbon dioxide throughout its life. If the rearing room contains carbon dioxide and other gases the worm vomits its intestinal juice at 12 to 13 per cent carbon dioxide in the air and dies at a 15 per cent concentration. Hence, during rearing ventilation facilities or open air rearing are mandatory.

Light

Light exerts a major influence on the growth of silkworms. Silkworms prefer light of 15 to 30 luxes and avoid string light and darkness. The standard recommendation is that silkworms are reared under dim light during the day and in darkness at night. The method of rearing under 16hr photoperiod alternating with 8hr darkness is reported to give systematic and uniform growth of larvae and the same schedule is followed.

Spacing

Larval growth will be very fast in the silk worms during the early stages. Too much crowd in the rearing tray results in increase in humidity, heat, fermentation of litter which results in un-hygienic conditions, wastage of leaf and under development of silkworm. The number of trays and space required for each instar will increase.

Stage Silkworm	*No. of Trays Required for 100 DFL's (Diameter of tray 3 ½)*		*Space Required (Bi and Multi x Bi hybrids)*	
	Beginning	*End of stage*	*Begin*	*End*
1st instar	2	2	4 to 15 Sq.ft	
2nd instar	2	5-6	15 to 45 Sq.ft	
3rd instar	5-6	10-12	46 to 90 Sq.ft	
4th instar	10-12	20	91 to 180 Sq.ft.	
5th instar	20	40	181 to 360 Sq.ft	

Spacing should be given simultaneously at the time of bed cleaning or feeding and care should be taken not to disturb the worms very frequently.

Feeding of Leaf

The leaf is finely chopped as per the recommended size and sprinkled on the larvae and after half an hour the newly hatched larvae slowly crawl on to mulberry leaf. Chopped leaf should be preserved from drying and loss of moisture content. For this purpose it should be collected in a basin covered with paraffin paper to protect from loss of moisture content.

First feeding to the newly hatched larvae should be given at 9 A.M. in the morning. During the 1st instar depending upon the silkworm strain and other factors (environmental conditions and leaf quality) the required mulberry leaf per 100 DFL will be around 2-2.5 kgs. The 1st instar stays for 3-31/2 days and undergo moult. The moulting period last for about a day. This will depend upon the environmental conditions.

Gap should be maintained for feeding the worms before undergoing moult and immediately after moult. Generally two hours gap is given. Also over feeding should be avoided, especially before undergoing moult which will result in wastage of leaf. Not only the nutritive value but also the number of feeds in each instar plays a major role in the cocoon built. Three to four feeds are given to the silkworm. In a day the last feeding *i.e.*, during the night should be a little more since the duration for the next feeding will be longer. During moulting period no feeding should be given. Before settling to moult and immediately after moult, first feeding should not be heavy. Feeding tender leaves to young age worms is essential. As the larval growth advances the mature leaf can be fed.

Maximum leaf is consumed during 5th instar only. During this stage only the maximum growth of silk gland can be noticed. It is estimated that 50 per cent of the total weight will be increased in the 5th instar itself. The feeding schedule is:

Sl.No.	1st Feeding	2nd Feeding	3rd Feeding	4th Feeding	5th Feeding
1	6 A.M.–6.30 A.M	11–11.30 A.M	3–3.30 P.M	7–7.30 P.M	
2	7 A.M.–7.30 A.M	11–11.30 A.M	2–2.30 P.M	5–5.30 P.M	
3	6 A.M.–6.30 A.M	10–10.30 A.M	2–2.30 P.M	6–6.30 P.M	10–10.30 P.M.

More number of feedings are given during summer since moisture content in the leaf will not be sufficient during summer and leaves will dry very easily.

Leaf Quality

Quality of mulberry leaf in addition to proper maintenance of temperature/ humidity plays significant role in the development of healthy silkworm.

Leaf quality plays an important role in the production of quality cocoons. The young age worms are fed with tender, succulent leaves which contain sugar, less amount of fibre, starch but, high moisture and protein that are suitable for chawki worms. Hence, the top tender leaves of mulberry plant are used to feed the young age worms. The late age worms are fed with matured leufs available at the bottom of the mulberry trees.

The leaf during its feeding time should not contain dust particles or water droplets on the surface of leaf. Disease attacked, wrinkled, weak leaves, poor quality leaves should not be fed to the worms. Feeding bottom matured leaves to chawki worms, tender leaves to mature worms should be avoided.

Preservation of Leaf

Harvested leaf should be transported in wet gunny bags or baskets made up of bamboo. Such leaf should be preserved in a separate room or in a corner of rearing room or in specially designed leaf preservation chamber made up of wood with sufficient number of ventilators.

The leaf chamber should contain a good number of ventilators covered with wet gunny cloth. This will help in maintaining leaf quality. Otherwise leaf should be scattered on the floor in a single layer so that the water droplets on surface of leaf will be evaporated. Leaf should never be preserved on heaps. This will increase the humidity. Leaf scattered on floor should be frequently sprinkled with water and tilted in summer.

Leaf Requirement

Silkworm attains nearly 10,000 times of weight starting from hatching to spinning stage, therefore, feeding quality leaf plays an important role in the development of silkworm.

Sl.No.	*Silkworm Stage*	*Quantity of Mulberry Leaf Required (approx.) (Kgs)*
1	1st Instar	2-4 (0.3-0.5 per cent)
2	2nd Instar	4-8(0.6-1.0 per cent)
3	3rd Instar	30-40 (4.2-5.1 per cent)
4	4th Instar	80-90(11.0-11.4 per cent)
5	5th Instar	600-650 (83.6-83.1 per cent)

The requirement of leaf will differ from race to race. The consumption of leaf also varies with season. During summer the quantity of leaf required will be a little higher since, drying of leaf will be faster hence additional feeding may be required.

11.16 Silkworm Nutrition

- ☆ In silkworm there is significant interrelationship between nutritional efficiency and silk production
- ☆ It affects the fibroin and sericin content
- ☆ Food conversion efficiency of silkworm is considered to be important physiological criteria for evaluating the superiority of the breeds or hybrids.

I. Important Nutrients Required by the Silkworm

a. Proteins

These are macromolecular polymers comprise of amino acids as base units and they also contain carbon, hydrogen, oxygen sulphur, phosphorus and nitrogen.

Five groups of L-amino acids required by the silkworms.

Essential amino acids (Group 1)

Arginine, Histidine, Isoleucine, leucine, lysine, methionine
Phenylalanine, threonine, tryptophan and valine.

Essential amino acids (Group 2)

Asparticacid and Glutamic acid.

Semi essential amino acid

Proline

Non-essential amino acids (Group 1)

Alanine, Glycine and Serine

Non-essential amino acids (Group 2)

Cystine and Tyrosine

b. Carbohydrates

Carbohydrates are principle organic compounds contribute mainly to structure and function of insect tissues and they all found in nucleus, cytoplasm, cell membrane as well as extra cellular fluid (Heamolymph) and other supportive tissues.

Important Carbohydrates Required by the Silkworm

Xylose	Sucrose
Arabinose	Maltose
Rhamnose	Lactose
Ribose	Cellobiose
Galactore	Trehalose

c. Lipids

Lipids are very important heterogeneous group of organic compound having relative insolubility in water and high solubility in non power solvents such as chloroform, acetone and benzene.

The lipid content of the dry matter of mulberry leaves ranges from 3 to 6 per cent the co-efficient of digestibility of lipids of mulberry leaves is reported to be 58.5 per cent on average from the 1st to the 5th instars. And silkworm accumulates more lipids in its body then the amount digested.

Sterols present in mulberry leaf

- ☆ Beta – sitosterol
- ☆ Campesterol
- ☆ Stigma sterol

Fatty acids required by the silkworm

- ☆ Myristate linoleate
- ☆ Palmitate linolenate
- ☆ Stearate arachidate
- ☆ Oleate

Lipid is another form of energy stored present in the silkworm

Lipid is a component of sub cellular membraneFor a formation of membrane, this controls the movement of gases, liquid and solid content of metabolized food in to the cell.

Precursors of vitellogenetic hormone ecdysone

The linolenate and linoleate were most effective in silkworm growth. Deletion of both resulted in a marked depression of growth. Based on the studies it may be concluded that these two polyunsaturated fatty acids are essential for silkworm although other fatty acids posses some nutritive value.

Constituent of cuticular wax.

Constituent of lipoprotein carrier molecule

Beta sitosterol act as a phagostimulant

It present in mulberry leaf in adequate quantity.

A minimum quantity of cholesterol in food results in normal growth

d. Vitamins

Vitamins present in mulberry leaf. Almost all the vitamins are required for silkworm growth among these water soluble vitamins play a major role than the fat soluble vitamins. By viewing the table it is clear that for a healthy cocoon crop, mulberry leaf rich in vitamin content are required.

Vitamins are organic compounds, which are required in small quantities for the normal growth and maintenance of life.

1. Vitamins functions as cofactor of enzymes
2. Vitamins are required as lipogen substrates.
3. Vitamin A is essential for growth, protein and mucopolysaccharide synthesis, fertility, and carbohydrate and DNA metabolism

4. Vitamin D helps in calcium and phosphate absorption and calcium metabolism
5. Vitamin E is necessary for fertility, normal functioning of muscles and to prevent oxidation of vitamin A, fatty acids and sulphur
6. Beta carotene and alpha tocopherol and slightly effective in increasing the fecundity in silk moth.
7. Vitamin K plays a vital role in coagulation of haemolymph on the wound surface of silkworms.

Classification of Vitamins Based on Solubility

1. Water Soluble Vitamins

- ☆ Vitamins B:
- ☆ B1 (Thiamine)
- ☆ B2 (Riboflavin)
- ☆ B3 (Pantothenic acid)
- ☆ B4 (Niacin)
- ☆ B6 (Pyridoxine)
- ☆ B7 (Biotine)
- ☆ B9 (Folicacid)
- ☆ B12 (Cobalamine and inositol)
- ☆ Vitamin C (Ascorbic acid)

2. Fat Soluble Vitamins

- ☆ Vitamin A (Retionol)
- ☆ Vitamin D (Calciferol)
- ☆ Vitamin E (Tocopherol or antisterility vitamin)

e. Minerals

Minerals required by the silkworm and they reported from the point of silkworm nutrition, mineral elements are indispensable to enhance the larval development and cocoon yields, but the mode of complementation, right dosage, time of application number of sprays have all to examined afresh to secure maximum benefit from such nutrition. If suitable dose and treatment are advocable and practiced at the reares level, this can give the much needed boost to overall production of the silk and seed.

Minerals are the inorganic part of the food plays on important role for successful development of insect.

The important minerals required by the silkworm are potassium, phosphorus, and magnesium, zinc. The extent to which the ingested minerals are retained as body substances was reported to be approximately 25 per cent throughout the whole instars. Concentrations of K is consistently higher than that of Na in the blood of silkworm and even in the whole body.

Quantitative Nutritional Requirement of the Silkworm

Crude protein	63 per cent
Fat	10.6 per cent
Ash	7.4 per cent
Carbohydrates	18.2 per cent
Calcium	0.34 per cent
Phosphorus	0.8 per cent
Ca: P ratio	1:2.35

Quantity of different nutrients required by the silkworm and amount present in mulberry leaf.

Amino Acids

The amount present in mulberry leaf satisfies the quantity required by the silkworm.

f. Minerals

1. Cobalt

a. Cobalt exerts favorable effect on the growth of silkworm
b. It enhanced the concentration of some other trace elements and hydrolytic enzymes of silkworm
c. It increases the larval weight, shell weight and oviposition rate.

2. Magnesium

a. It accelerates the growth of silkworm through orientation of physiological activities.
b. It reduces the larval duration, increases the oviposition rate in the adult.
c. It increases the silk content in the cocoon

3. Potassium

a. It increases the larval body weight and shell weight in silkworm
b. It enhances the silk gland weight, oviposition rate, proteins and RNA content of silk gland
c. It also increases the fecundity

4. *Iron*

a. It improves the cocoon weight, cocoon size and rearing percentage.

b. It influences the cocoon yield and shell percentage.

5. *Zinc*

a. It increases the pupal weight and filament length.

b. Zinc is important for reproduction

c. It is much useful in grainage centers for increasing seed production.

6. *Phosphorus*

It increases the larval growth, economic characters and fecundity.

7. *Calcium*

It increases the larval growth, cocoon characters like cocoon weight, shell weight, cocoon/shell ratio and silk proteins (Fibroin and Sericin)

Deficiency of Nutrients in Silkworm Growth and Development

1. Deficiency of Lipid

In the absence of linolenic and linoleic acids in the diet during the 4th and 5th larval instars, malformed wings are developed when the adults emerged

The hatchability of the eggs laid by such adults was greatly decreased and death of eggs occurred during the embryonic stage.

2. Vitamin Deficiency

a. Vitamin B

1. Deficiency paves the way to cent per cent mortality in I and II instars.
2. Retardation of growth and development in late age worms.
3. Choline deficiency leads to inability for moulting.

b. Vitamin C

Deficiency resulted in the high mortality of late age worms.

c. Vitamin A

i) Retarded growth

ii) Loss in weight

d. Vitamin D

Retardation of growth due to non accumulation of minerals like calcium and phosphate.

11.17 Artificial Diets for Mulberry Silkworm, *Bombyx mori* L.

The application of artificial diets for silkworm nutrition was reported in 1960's. The artificial diet used for silkworm rearing during that period contained dried pulverized mulberry leaves as one of the main ingredient and the composition was rather simple because no information on the nutritional requirements of this insect was known by then. When silkworms were reared on these diets larval growth and development were retarded, small cocoon appeared and adults laid only a small number of eggs. Subsequently, diets containing 10 per cent mulberry leaf powder with starch, sucrose, soyabean powder and cellulose powder were prepared. Later on it was shown that increase in the concentration of cellulose powder upto 48 per cent improved the feeding activity. Further the diets were substituted with casein + glutien or soyabean powder + casein which improved the acceptability of the diet.

Based on the composition of the diet, three artificial diets have been developed.

1. Amino acid diet – contains aminoacids as a sole source of nitrogen.
2. Diet containing mulberry leaf powder.
3. Diet with out mulberry leaf powder/semi-synthetic diet.

1. Composition of Amino Acid Diet

Substance	*Per cent in Dry Diet*
Potato starch	7.50
Sucrose	12.50
Amino acid mixture	20.00
Soybean oil, refined	3.00
β-sitosterol	0.50
Werson's salt mixture	4.50
Ascorbic acid	2.00
Citric acid	0.50
Cellulose powder	34.20
Morin	0.30
Agar	15.00
Total	100.00
Vitamin-B-mixture	Traces
Distilled water	3 ml/g dry weight

All the constituents except Vitamin-B-mixture is added into distilled water and sterilized at 121°C temperature and 15 psi for 15 minutes, while Vitamin-B-mixture is added into cooled molten (45°C) diet.

Amino acids: alanine, arginine, K-aspartate, cystine, Na-glutamate, glycine, histidine-HCl, isoleucine, leucine, lysine-HCl, methionine, phenylalanine, proline, serine, threonine, trypotophan, tyrosine and valine.

Vitamin B – Mixture: biotin, choline, folic acid, inositol, niacin, pantothenate, pyridoxin, riboflavin and thiamin.

Amino acid diets sustains larval growth to some extent and the larva can spin the cocoons. This diet is used for fundamental nutritional studies, but not for commercial use.

2. Composition of Artificial Diet Containing Mulberry Leaf Powder

Substance	*Diet for 1st–4th Instar (g)*	*Diet for 5th Instar (g)*
Mulberry leaf powder	25.00	25.00
Soybean oil	1.50	3.00
Defatted soybean meal	36.00	45.00
Cholesterol	0.20	0.20
Citric acid	4.00	4.00
Ascorbic acid	2.00	2.00
Sorbic acid	0.20	0.20
Agar	7.50	5.00
Salt mixture	3.00	3.00
Glucose	8.00	10.00
Potato starch	7.50	15.00
Cellulose powder	20.80	-
Vitamin B mixture	Traces	Traces
Antiseptic	Traces	Traces
Total	115.70	112.40
Distilled water	300 ml	220 ml

Preparation of this diet is as that of amino acid diet, wherein, Vitamin-B-mixture and antiseptic are added into cooled molten diet. This diet is used for commercial rearing of silkworm.

3. Composition of Semi-Synthetic Diet

Substance	*Dry Diet (g)*		
	Starter (I instar)	*Grower (II-IV instar)*	*Producer (III instar)*
Potato starch	10.00	10.00	20.00
Sucrose	10.00	10.00	-
Glucose	-	-	12.00

Substance	*Dry Diet (g)*		
	Starter (I instar)	*Grower (II-IV instar)*	*Producer (III instar)*
Soybean meal, defatted	30.00	40.00	60.00
Soybean oil, refined	3.00	3.00	3.00
β-sitosterol	0.50	0.50	0.50
Salt mixture	3.50	3.50	2.00
K_2HPO_4	1.00	1.00	-
Ascorbic acid	2.00	2.00	2.00
Cellulose powder	34.00	34.00	-
Agar	15.00	15.00	5.00
Citric acid	0.50	0.50	0.50
Sorbic acid	0.20	0.10	-
Total	109.90	119.80	105.20
Vitamin B mixture	Traces	Traces	Traces
Antiseptic	Traces	Traces	Traces
Distilled water	300 ml	300 ml	260 ml

The preparation of this diet is similar to that of artificial diet containing mulberry leaf powder.

In Japan, diets were earlier prepared to rear only chawki worms, but now diets are available for rearing both chawki and late age silkworms. However, at present artificial diets are supplied only for I – III instar and mulberry leaves should be provided to IV and V instar in a large rearings.

Soyameal

Soya bean meal is commonly added to the artificial diets as a main source of protein because of its high nutritive value as indicated from the nutritional research conducted on defatted soybean meal and various proteins. Nutritional research has proved that dietary levels of meal influenced the larval growth as well as cocoon yield and thus recommendations have been on their practical implication in sericulture in advanced countries. Owing to the high cost of artificial food, introduction of these techniques in India is very limited. Hence an economically cheaper technique which could still improve cocoon character and production is needed. Fortification of mulberry leaves with soybean products fulfilled the above need. Quantitative and qualitative supplementation of protein plays a critical role in influencing the growth and development of the insect. Elevation of dietary protein to an optional level and supplementation of low nutritive proteins with their limiting amino acids have found to accelerate the growth of the silkworm.

These twin objectives were fulfilled by protein supplementation with soya products. soya protein is known to increase the protein and amino acid content in the larval haemolymph of the silkworm *B.mori.* andpromote growth and to improve the economic character of the silkworm.

Supplementation of Flour Diets

Three flour diets *viz.*, rice, wheat and ragi were fed to silkworms with mulberry leaf to evaluate the cocoon characters. Supplementation of flour diets increased the larval weight, silk gland weight and commercial cocoon characters. These are also known as fortification agents or supplementary nutrients when added to the normal diet. They enriched the nutritional value of the diet, making it more useful from the nutritional point of view.

Most popular and commercially used artificial diets in Japan are

1. Silkmate
2. Mayomisilk
3. Morus
4. Vitasilk *etc.*

Under Indian conditions, it has been experimentally proved at CSR and TI, Mysore, that diets can be used in silkworm rearing.

Advantages

1. Balanced nutrition can be provided to silkworms, according to their age.
2. The growth and development of silkworms is uniform.
3. Maintenance of aseptic conditions helps in prevention of pathogens.
4. Ideal method for studying nutritional effect on silkworms.

Limitations

1. The cost of rearing silkworm on artificial diets is 4 times more than rearing on mulberry leaves.
2. Maintenance of aseptic condition in the farmer's field is a bottle neck.
3. Diets are easily contaminated by bacteria.
4. Require special conditions to prepare and store the artificial diets.

Use of Juvenile Hormone Analogs on Silkworm Rearing

Some chemicals analogous to insect hormone, such as the synthesized juvenile hormone and moulting hormone (ecdysone), which is extracted from some plants, have been used in silkworm rearing. **Juvenile hormone:** There are many substances analogous to the juvenile hormone; amongst those which are produced and used in china are "JH- 734", "JH -738", "JH-512" and "JH-515".

The use of these substances in sericulture production results in physiological changes in the worms, chiefly the lengthening of period of the fifth instar. That is to say, the time required for eating and the amount of mulberry leaves eaten would be both increased. Since the amount of nutrients accumulated and assimilated would be increased, body would increase and the weight of silk substances would increase as well. The final goal of raising the production of silk could be attained. Practical operation process that the production of silk can be increased by 8-10 present for the bivoltine variety and 10-15 percent for the multivoltine variety if this kind of hormone is used. The technical skill needed to use juvenile is as follows.

Dosage and Concentration

In using this hormone, it is important to choose the dosage that is safe to the larvae and that will increase silk production.

Time for Spraying

The time to spray juvenile hormone is a bit earlier than the middle stage of the fifth instar. It must be determined delicately according to the variety of the silkworm and the temperature and humidity. Since the fifth instar last longer in the bivoltine than in the multivoltine, the spring and late autumn crop, spraying the hormone should be done late. If the atmospheric temperature is too low and in hot season, it is favourable to spray the hormone 42-60 hours after the first feeding.

11.18 Cocoon Harvesting

Mounted silkworm can complete cocooning in about two days. It becomes pupa in another two days. At the beginning of pupation, the skin of pupae is soft fragile and liable to be wounded easily. If cocoons are harvested at this stage, the interior layer of cocoon shell is soiled which affect, the reliability. It takes 2 or 3 days after pupation for the skin to harden. Hence, the cocoon should be harvested only on the 6th day after mounting. The temperature should be 25-27°C. therefore, it is safe to conduct cocoon harvesting after hardening of the pupal skin is tested by shaking the cocoons gently.

What are measures to be taken before harvesting the cocoons?

1. Cocoons harvested method depends on the kind of mountage used.
2. Before harvesting all diseased and unspun larvae should be removed.
3. Remove the faecal matter and unwanted cocoons.

Harvesting of coons are two types

1. Manual
2. Mechanical

Manual means it is done by hand that is labour requirement and individual cocoon harvesting. This is practiced in developing countries like India.

Mechanical this is done by machine in developed countries and it is power requirement. This id carried out in rotary mountage.

Transportation

This process is done after the cocoon harvested. Transportation should be done in cooler hours that are morning or evening and separate vehicle should be provided to avoid the damage for cocoons.

11.19 Natural Enemies of Silkworms

Silkworm throughout the world is prey to a number of invertebrates and vertebrate pests during rearing and mounting as well as in cocoon storage rooms. Unless precautionary measures are taken against these pests, considerable loss in yield is encountered. The major invertebrate pests are ants, Uzi files and straw mite while lizards, birds and rats are vertebrate's predators.

Uzi (*Exorista bombycis*)

The uzi is also undergoes complete metamorphosis and has four stages in its life cycle, namely egg, larva, pupa and pupa and adult, of which only, larva is an internal parasite of the silkworm larva.

Symptoms

Uzi fly infestation can be inferred by observation of the egg attached to the silkworm body that spot becomes black at the site of infestation. Uzi infestation stops feeding become lethargic and lose weight continuously as the parasitic maggots utilize all nutrients stored in the fat bodies. It eats all the tissues in body. The body of the silkworm becomes lusterless.

Straw Itchmite (*Pyemotus ventrocosus*)

This is not an insect pest but belongs to group of mites. This mite is encountered at the time of rethatched of the roof of silkworm rearing house or brought with building materials.

Symptoms

The body surface of silkworm stuck with this mite develops a few black specks. The infested silkworm loses appetite, becomes inactive and has deficiency in excreting. It takes time to pass the excreta and frequently the excreta are attached bead like to the anus. Silkworms vomit yellowish green fluid and excrete black fluid. Irregular or decrease in pulse of dorsal vessel. The motes derive nutrition from the silkworm and give out a toxic substance which kill silkworm.

Rats (*Rattus rattus*)

Rats eat grown up worms in rearing trays and ripe worms on mountages. During spinning they cut the cocoons and fed on spinning worms. During this process, the whole cocoon is damages and become unfit for reeling.

Ants

Ants attack the silkworm in the trays, spinning larvae and cocoon on mountages.

Squirrels

These mammalian predators eat the worms as such and also pupae after biting open the cocoons.

Lizards

They are frequent in the rearing house and often pick up the young age worms and eat them, thus causing considerable damage.

Earwing (*Labia arachidis*)

The earwigs known to attack the moths in grainage. This attack female's moths. It holds the abdomen of moths with help of mandibles and pinch the same with aid of forcep-like cerci. A black scar develops on the moth at the point of damage from which black body fluid oozes out. It affects on oviposition and lays less and dead eggs. It feeds on body skin of the moth.

Birds

Birds such as crows and sparrows pick up the silkworms when the mountages are kept out-of-doors at the time of spinning.

Chapter 12

Seed Technology

12.1 Seed Technology: An Introduction

Egg is the backbone of silk industry. Healthy egg production is indeed the primary requisite for conducive growth of the entire silk industry. Timely supply of superior quality of silkworm seed can alone sustain sericulture as a commercial crop in competition with other cash crops. A key factor for success of Sericulture industry is not only augmentation of silkworm food plants but timely supply of superior quality of silkworm seed to the farmers should also be ensured. The quality of silkworm seed may be defined as to the one where the layings are entirely free from diseases, has more nos., of viable eggs, gives uniform hatching and assures a stable crop. The government grainages produce only 10 per cent of total requirement. Remaining 90 per cent seeds are produced by the farmers themselves for their own requirement without resorting to scientific procedures. This practice gives scope for spread of dreaded silkworm disease called pebrine as microscopic examination of mother moth is never resorted or recommended package and practices are not followed by the farmers.

To produce good quality eggs there must be a sound seed organization with proper elite seed multiplication network to produce large quantities of quality commercial seed. The muga raw silk production has not shown significant growth and has remained fluctuating. The fundamental reason behind the poor production is shortage of quality muga seed (eggs) for commercial rearing. Non- availability of quality seed required quantity and in the proper season, for conducting the commercial crop rearing has been attributed to be the major factor the decline of muga silk production. Unlike in mulberry sericulture, the seed is transacted in the form of seed cocoon in muga silk industry which may be the prime factor for the prevalence of pebrine disease in muga silkworm. The crop loss due to pebrine continues unabated and every rearer prepares his laying from the cocoons he chooses as seed cocoons; the mother moth after oviposition is not microscopically examined for pebrine and hence the seed prepared is not free of contamination. Thus, the muga seed prepared and used by the rearers is of poor quality. Even the departmental agencies supply only the seed cocoons to the reares, and majority of the rearers only prefer seed cocoons and they are very reluctant to accept microscopically examined layings, possibly they are not prepared to change their age-old traditional practice and adopt scientific technology. Even the seed cocoons are not readily available to the rearers.

Silkworm seeds are produced in the grainage and therefore grainage is the first and the most important aspect which needs special attention. Success of silkworm rearing largely depends on careful grainage operation. Systematic and methodological grainage not only minimize larval morality due to diseases but also results in vigorous progenies. The grainage operations consist of selection, storage and preservation of seed cocoons, egg production.

The CSB has a chain of Basic Seed Farms supplying basic seeds to the States. Its commercial seed production centers augment the efforts of the States in supplying commercial silkworm seed to farmers.

Silkworm Seed Production in India (Unit: Lakh dfls)

Particulars	*2019-20*		*2020-21*		*2021-22*	
	Target	*Achmnt.*	*Target*	*Achmnt.*	*Target*	*Achmnt.*
Mulberry	470.00	399.87	410.00	356.18	400.00	329.74
Tasar	51.17	55.53	52.77	47.37	51.40	47.46
Oak Tasar	1.48	0.44	0.576	0.50	0.138	0.053
Muga	5.65	5.71	5.86	5.72	6.463	6.20
Eri	6.30	6.64	6.00	6.48	6.00	6.45
Total	534.60	468.19	475.20	416.25	464.001	389.903

12.2 Grainages

Grainages are the synonymous terms used in seed production centres. Grainages are establishments where disease free and quality seeds are produced on scientific lines. The aim of a grainage is the production of quality seeds. Quality seed is the backbone of sericulture industry. Quality seed is a seed which is free from diseases, has more number of viable eggs, give good uniform hatching, produced from healthy and robust parents and assures a stable and successful cocoon crop.

Types of Grainages

1. Basic Seed Stations/Centres

These are the stations where the layings of the parental races are produced and managed by government agencies. The race breeding stations are maintained by the government. These centres multiply great grand parents or the grandparents (P4 or P3) of commercial Grainage seeds. These centres contain all pure races, andare aimed to maintain purityofthe races. Thesecentres supplythe basicor initial material for multiplication, to the breeding centres. Normally the seeds of multi-voltine races (pure or CB) are multipliedin three-tier system(P3) while exotic pure breed races of bi-voltine, rare multi- voltine have four-tier system(P4).These stations(P3 or P4) supply parental seed cocoons for rearing at foundation stock seeds (P2 Station) to produce P1 seed.

2. Industrial Seed Grainages or F-1 Centres

The industrial seed Grainages may be private or government owned. These Grainages are aimed to produce reproductive and industrial seed. A multiplication of pure breed races is done by seed cocoon rearers who in turn produce seed cocoons (P1). These seed cocoons are brought to the industrial seed producing Grainages to produce the hybrid seeds (F1) which are also called foundationhybrids.

Functions of Grainages

The aim of a grainage is the production of quality seeds. The silkworm eggs required for commercial rearing should be of high quality and free from diseases. To produce good quality seed there must be a good seed organization.

Silkworm seed is divided into two types *i.e.,*

a. Reproductive seed

b. Industrial seed.

a. Reproductive Seeds

These are used for producing the seed cocoons (parents of seed cocoons) which are required in large numbers for producing commercial seed. The purpose of these is for maintaining the racial purity which is difficult to rear, So special care must be taken by technical staff. The selling price of these is 30- 50 per cent more than reeling cocoons. Reproductive seeds are often multiplied in number in a series of breeding centres called breeding stations (P4 or P3, P2, and P1). In three or four stages in order to ensure that the racial characters are not diluted during the multiplication stages.

b. Industrial or Commercial Seed

There are specific hybrids between two or more pure lines of silkworm races and are reared by the sericulturists for producing cocoons on a commercial scale for reeling purpose. These are hybrid seeds, produced in special organizations called Grainages.

There are three aspects in seed production. They are

1. Breeder's stock.
2. Basic seed multiplication.
3. Industrial seed production.

The race breeding stations are maintained by the government. These centres multiply great grand parents or the grandparents (P4 or P3) of commercial Grainage seeds. These centres contain all pure races, and are aimed to maintain purity of the races. These centres supply the basicor initial material for multiplication, to the breeding centres.

Normally the seeds of multi-voltine races (pure or CB) are multiplied in three-tier system (P3) while exotic pure breed races of bi-voltine, rare multi- voltine have four-tier system(P4).The above said stations(P3 or P4) supply parental seed cocoons for rearing at foundation stock seeds(P2 Station).A multiplication of pure breed races are done by seed cocoon rearers who in turn produce seed cocoons. These seed cocoons are brought to the industrial seed producing Grainages (P1) to produce the hybrid seeds which are also called foundation hybrids.

12.3 Disinfection and Hygiene in Seed Production Units

In order to ensure success full grain age operation, the grainage hall and the equipments should be disinfected properly. The disinfection operation can be categorized in to two stages Washing and spraying and Fumigation.

- ✰ **Washing and Spraying:** The grain age hall and its equipment at first should be washed thoroughly with clean water and/or bleaching powder. After 24 hours they should be sprayed with 4 per cent formaldehyde solution and should be kept air tight for 48 hours
- ✰ **Fumigation:** After 48 hours of formalin spray, 35 to 40 per cent commercial formaldehyde solution should be boiled inside the grain age hall keeping it air tight so that the vapours of the formalin effectively disinfect the room. After 48 hours the hall should be opened for proper aeration and should be washed with clean water. For production of dfls and have good crop one should consider the following steps.

Disinfection before Grainage Activity

- ✰ Grainage rooms along with the equipments should be disinfected and kept ready before the arrival of seed cocoons
- ✰ Seal the grainage hall 5-7 days prior to operation.
- ✰ Drench the walls and other bigger appliances with 5 per cent bleaching powder solution.
- ✰ Dip smaller appliances in 5 per cent bleaching powder solution.

- ☆ Use high grade bleaching powder with more than 30 per cent chlorine content.
- ☆ Spray 2 per cent formaldehyde solution and 0.5 per cent slaked lime mixture @ 1 lit. per 2.5 sq.mtr. and leave the hall closed.
- ☆ Use disinfection mask, overcoat and hand gloves while disinfecting the grainage hall.
- ☆ Prepare fresh disinfectant by mixing 5 gm slaked lime in 1 litr. Of 2 per cent formalin solution.
- ☆ As formalin is more effective a above 20°C, carry out disinfection preferably on sunny days.
- ☆ During high humid conditions the hall may be fumigated with 5 per cent formaldehyde solution for 24 hours at least 3-4 days before and immediately after grainage operation.
- ☆ Complete the process of disinfection at least 3 days before consigning seed cocoons.

Hygiene

- ☆ Store seed cocoon in single layer in moth cages to avoid the rate od contamination Burn the dead and melted cocoons.
- ☆ Use hand glove, aprons and foot wears during grainages operation.
- ☆ Sprinkle bleaching powder and slaked lime mixture (100gm bleaching powder and 900 gm slake lime) at the entrance and around the grainage hall. @ 200 gm per sq.mtr.) 1-2 mtr. Around on alternate days.
- ☆ Disinfect the moth testing area regularly with 2 per cent formalin and 5 per cent bleaching powder solution.
- ☆ Dispose off the pierce cocoons as soon as possible.
- ☆ Dispose off crushed moths, papers *etc.* in soak pits after treating with 5 per cent bleaching powder solution or burn them.
- ☆ Make soak pits always away from the grainage hall.
- ☆ Wash hand with alkaline soap and then 2 per cent bleaching powder solution before entering the grainage hall.
- ☆ Disinfect footwear by steping on the mat soaked with 2 per cent formaldehyde solution.

12.4 Price Fixation for Seed Cocoons

The seed crop rearings and conducted during unfavourable seasons and hence the success of the seed crops is highly uncertain. However, once the cocoons are produced, they are sold to commercial rearers at very high price. Often the seed cocoon price depends on the place, demand and availability.

The seed cocoons are sold in numbers of hundreds, thousand *etc.* and prevailing rates vary from rupees four hundred to five hundred per thousand seed cocoons. It is estimated that the current requirement of seed cocoons is about five crores to produce one crore commercial disease free layings (DFLs) as five seed cocoons are required for production of one DFL.

12.5 Norms of Seed Cocoons Required to Produce Quality Bivoltine Hybrid Seed

Production of quality bivoltine hybrid seed can be achieved by using bivoltine parent seed cocoons with following parameters:

- ✰ Free from pebrine disease.
- ✰ Free from mixing of another race.
- ✰ Pupation rate of 80 per cent and above.
- ✰ Yield per 100 dfls is 45 kg and above.
- ✰ Average single cocoon weight of 1.55 g to 1.75 g *i.e.*, 570 to 650 cocoons per kg.

Production of quality bivoltine hybrid seed can be achieved by using bivoltine parent seed cocoons with following parameters:

- ✰ Free from pebrine disease.
- ✰ Free from mixing of another race.
- ✰ Pupation rate of 80 per cent and above.
- ✰ Yield per 100 dfls is 45 kg and above.

Production of quality multivoltine hybrid seed can be achieved by using multivoltine parent seed cocoons with following parameters:

- ✰ Free from pebrine disease
- ✰ Pupation rate of 90 per cent and above.
- ✰ Yield per 100 dfls is 42 kg and above.
- ✰ Average single cocoon weight of 1.00 g

12.6 Quality Assessment, Purchase and Transportation of Seed Cocoons

Parental seed cocoons are purchased at seed cocoon markets, while purchasing quality seed cocoons true to its parental breeds, uniformity in shape, size and as per number of cocoons per kilogram as a unit are selected. These cocoons are to be purchased after complete pupation (>80 per cent). It should be completely free from pebrine disease and yield of bivoltine cocoons should be more than 45kg/100dfls. After purchase it should be transported in cooler hours in plastic peforated crates or bamboo baskets in loosely packed condition. Pupal gut portion is taken out and subjected to microscopic examination. In case there is incidence of pebrine, the batch has to be rejected. Before the seed cocoons are arrived, it is very much important and essential to disinfect the rooms of the grainage.

Figure 1: Seed Cocoon Market.

Preliminary Examination/Selection and Sorting of Cocoons

In order to produce good quality and healthy eggs the seed cocoons used for the purpose must be of high quality and in good health and therefore, the seed cocoons after arriving at the grainages are subjected to rigid selection. In selection only sound and uniform cocoons conforming to the characteristics of the race are selected and defective and deformed cocoons, under sized and over sized cocoons, double, stained and dead cocoons, uzi infested *etc.*, are rejected.

Advance detection of pebrine disease, if any before the commencement of operation of each batch helps in averting great loss to the grainages. This is facilitated by investigations at three stages *i.e.*, Pupal test, Forced eclosion test and First day moth examination. Of the three tests, pupal test is ideal as it gives ample time to send the cocoons for reeling unit if they are infected by pebrine.

Pupa Test

In this test, the gut of pupa is more reliable for pebrine detection than the entire pupa. For this purpose, the pupa is cut ventrally just below the wing bud by a scissor by holding the pupa between thumb and for finger in left hand. After cutting the pupa is pressed gently. The mid gut oozes out as a brown body from the cut portion. This mid gut is collected and crushed with few drops of potassium hydroxide in a moth crushing set. The fluid is taken on the slide and examine under the microscope with 600 X magnification. If the stock is suffering from pebrine, the entire batch of cocoon is rejected and sent to market. Such cocoons should never be used for preparation of silkworm seed under any circumstances.

Preservation of Seed Cocoons

The cocoons are preserved in single layer in well ventilated rooms under natural light and dark conditions. Exhaust fans in cocoon preservation and moth emergence rooms are essential to expel foul gases and dust. Different component races are to be preserved in separate rooms.

Sex Separation at Pupal Stage

In case of pupal stage sexing, the cocoons are deflossed and cut at one end to remove the pupa for determination of sex. Cocoon cutting improves eclosion percentage in case of bivoltine cocoon, where the shell is hard and compact. Male and female pupae are

Male	*Female*
Smaller in size	Large in size
Narrow pointed abdomen	Broad abdomen
Small dot like mark on the ventral side near the top demarcation line of 9th abdominal segment	'X' mark seen on the ventral side of 8th abdominal segment

Synchronization of Moth Emergence

Moths of different races are made to emerge simultaneously on the same day and at the same time, so that the male and female moths are readily available for Hybridization. This phenomenon is referred as **"Synchronization"**. The synchronization process is to be planned from the brushing time of the parent races or cold preserved based on emergence time.

To prepare hybrid eggs, some adjustments are made so that the moths of different races emerge at the same time. This is done by selecting cocoons which spin on different dates, depending on the number of days required for emergence. If it cannot be adjusted for any reasons, the cocoons of the early emerging races have to be refrigerated at 5- 10°C for 2- 4 days. The refrigeration should be done preferably on the 7th or 8th day after spinning. The cocoons which are about to emerge should not be refrigerated. If refrigeration is not done in the cocoon stage, synchronization can be achieved by refrigeration of moths which emerge earlier.

Male moths can be refrigerated at 5°C for 7 days. Female moths can be refrigerated at 5°C for 2- 3 days. However, refrigeration of female moths is not advisable as it increases the percentage of unfertilized eggs and poor layings.

Moth Emergence

Generally moths emerge out of the cocoon after 9-14 days of spinning. But the pupal age varies according to races and seasons. Multi-voltine races emerge earlier than other races. The expected time of moth emergence can be predicted by observing the pupal and cocoon characters. A day before becoming moths, the pupae appear soft, dark in colour, with loose pupal skin, prominent eye formation, wing parts, appendages appear. A day before emergence a distinct sound is heard in the seed cocoon room, due to transformation of pupae into moths, which are trying to rupture the pupal skin to come out.

It requires a specific environmental condition for cocoon or pupae before emergence of moths. These are two be kept in total darkness. On the expected day of emergence, bright lights are illuminated in the early hours of morning at 4- 6 a.m. At the time of emergence temperature of about 25°C with humidity 70- 80 per cent should be maintained. Moths start emerging from 6-8 a.m. which are picked and kept in a tray for urination. Then healthy moths are are taken for coupling or for preservation. The emergence of moths decreases as the day light intensity increases. Moths do not emerge after 8.30 a.m. Light is provided only in the subsequent days of emergence. Thus room is once again made dark and light is provided only in the subsequent days of emergence. The emergence of moths lasts for 3-4 days only. Sometimes due to mistakes in sex separation moths of other sex mix and emerged moths mate. Such pairs are rejected.

Figure 2: Silk Moth Emergence.

While emergence the moth ruptures the pupal skin and secretes an alkaline juice on one end of the cocoon and makes it wet. Then moth using its legs and head, wriggles out of the cocoon. After emergence, moths stretch its wings for about 5 min. After passing urine especially male moths are seen moving actively in the tray.

Coupling/Mating/Pairing

After emergence of moths, the males and females are picked and allowed to mate as per the desired combinations. This is known as Pairing.

The female moths are collected in the tray into which the desired male moths are left over the female gently. The moths pair within 5-10 minutes. The left over male moths are collected into another tray. If the males available in excess, they should be preserved at 5oC for later use.

Each paired moths are kept in cellules arranged in a tray. All the trays are kept in dark room for about 3-4 hrs with a temperature of 25°C and 75-80 per cent humidity.

Depairing

Depairing is the process of separating the moths after 3 hours of mating. The moths are separated by holding the female lightly and moving the male sideways using the fingers. This facilitates easy separation without injury to the female reproductive organs. The trays in which the female moths are kept gently tapped to induce the moths to urinate.

Second Pairing

Properly preserved male moths pair 8 times, But with increase frequency of mating fertilization decreases rate of egg fertilization. Thus male moths can be used in pairing for 3 times. The male moths are preserved at 5°C temperature for about 4–5 days. Prior to second pairing the male moths preserved are released at room temperature for 5–10 minutes. Second pairing duration for about 4 hours with a temperature of 25°C and 80 percent humidity.

Oviposition

This egg laying process is known as oviposition. The oviposition room is kept dark by closing the doors and windows. At the time of oviposition a temperature of 23–25°C and humidity 80 per cent should be maintained for egg laying. If the humidity is less than 80 per cent the gummy substance secreted by the female dries up on the ovipositor and obstructs egg laying. This results in less number of egg laying, because most of the eggs remain in the body.

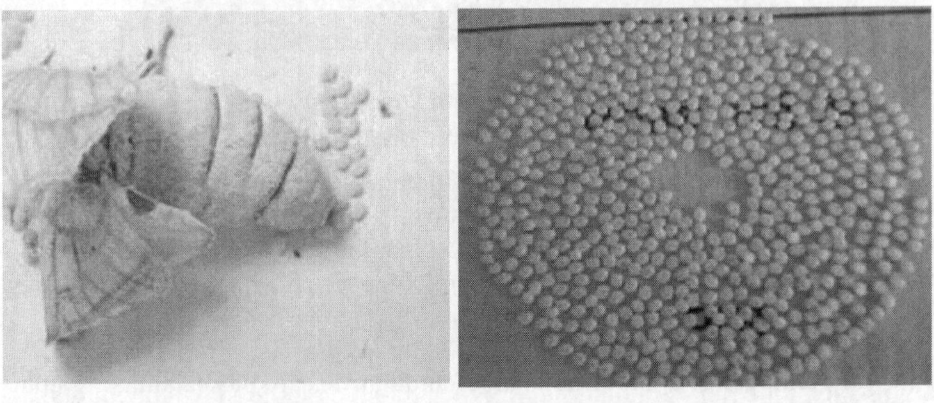

Oviposition

12.7 Egg Production Methods

Sheet Egg Production

Silkworm eggs are prepared in different forms like loose eggs or eggs on cards. In case of eggs on cards the eggs laid by each moth are isolated which facilitates individual moth examination and elimination of eggs laid by pebrine infected moths. The Non-hibernating eggs are generally of multivoltine or multi x bivoltine breeds. They hatch generally on ninth day itself because of high temperature.

Procedure

1. Preserve the seed cocoons of different breeds separately for moth emergence
2. Collect the male and female moths of different breeds separately
3. Allow the moths of desired combinations for mating for about 4 hours.
4. Decouple the pairs and allow for urination
5. Spread the egg card and transfer the mated female ones in each square and cover with cellules
6. Allow the moths to lay eggs for 24 hours
7. Collect the oviposited females for mother moth examination
8. Reject the eggs laid by infected moths
9. Wash eggs in cold water thoroughly and dry under shade

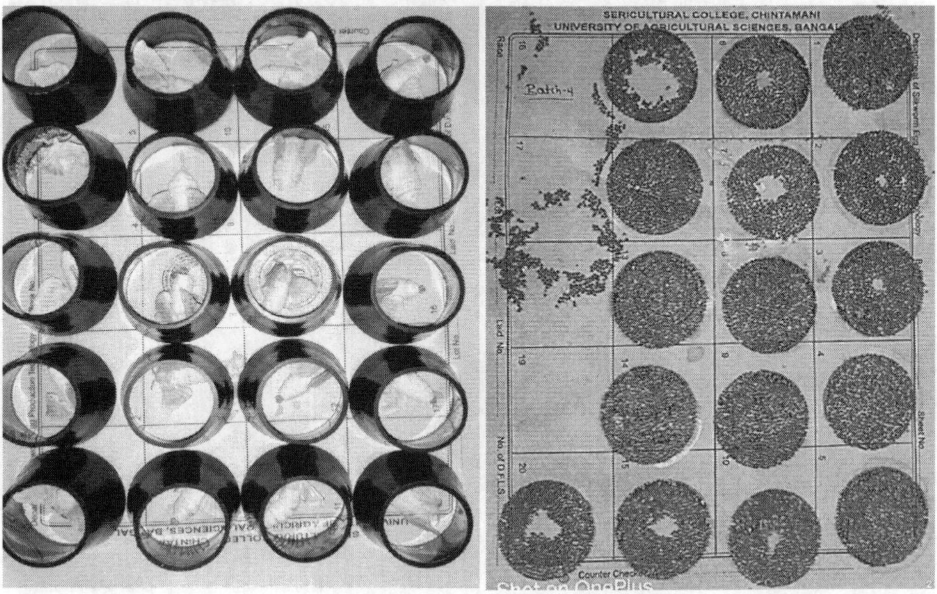

Sheet Egg Production.

Loose Egg Production

Among the different methods of silkworm egg production of eggs on cards/ sheet or in the form of loose grains are the best methods. Reproductive seed is usually handled in the sheet form. However, the industrial seeds are prepared both in the sheet form as well as in loose forms. By virtue of the advantages associated with loose egg production, in countries like China and Japan, commercial hybrids are invariably produced in loose form. On the contrary, in India, the very concept in rather new and is gaining momentum only in the recent years.

Preparation of Starched Egg Sheets

1. Take 40 to 50 g of maida in one liter of water and boil it to become a paste.
2. Smear the paste uniformly in a thin layer on craft paper sheet and allow it to dry
3. Store the well dried starched sheet in dry place to prevent attack of mould

12.9 Oviposition

Spread the starched egg sheets in the oviposition tray. The decoupled female moths are stimulated for urination and transferred on to the starched egg sheet. Spread the female moths uniformly at the rate of 40 to 50 (Bivoltine) and 50 to 60 (Multivoltine) per square ft in the oviposition tray. Keep the trays inside the oviposition stand and allow the moths to oviposit under dark condition for 24 hours. Maintain the temperature at 25°C +/- 1°C and RH of 75 +/- 5 per cent

Smearing Maida Paste on Egg Card

Loose Lgg Production

Collection of Eggs

Conduct mother moth examination as per standard procedure. Soak the disease free egg sheets in water for 20 to 30 minutes to soften the gum. Gently dislodge the eggs with hand. Collect the eggs in a double layered nylon mesh bags.

Degumming of Eggs

Prepare 0.4 per cent bleaching powder solution and allow it to settle. Soak the eggs in this solution for about 10 to 15 minutes to dissolve the gum. Consequently wash the eggs thoroughly in running water and dry. Follow the acid treatment for bivoltine eggs.

Packing of Loose Eggs

Collect the dried eggs, weigh one gram of eggs on electronic balance and count the number. Accordingly, calculate the weight of eggs to be packed. Insert a funnel into the slit of the loose egg box, pour the weighed eggs through the funnel into the box. Seal the opening with sticker. Label the box indicating the details like name of the grainage, breed, quantity, laid on date, released date and probable date of hatching and incubate the boxes with eggs. Usually each box contains 50 DFLs (20,000 eggs), which may weigh from 12.0 to 13.0g.

Advantages of Loose Eggs

- ☆ Easy for mass egg production, handling, transportation and incubation. Facilitates elimination of poor quality and unfertilized eggs before supply.
- ☆ Surface sterilization and hydrochloric acid treatment is more effective and uniform.
- ☆ More accurate than sheet eggs for comparison of scientific data like individual yields, race-wise, yields, region and season-wise yields.

Difference between the Sheet Egg and Loose Egg Production

Sheet Egg	*Loose Egg Production*
Ordinary kraft sheets without starch coating are used for egg laying.	Ordinary kraft sheets coated with starch are used for egg laying.
Kraft sheet used for egg laying is of 22.5 cm x 28 cm in size.	Starch coated kraft sheet used for egg layings is of 65 cm x 95 cm in size.
On each sheet 20 moths are left to lay eggs.	On each sheet 250 to 300 moths are left to lay eggs
Cellules are placed over the moths to restrict the egg laying area.	Cellules are not used here

Brine Treatment for Loose Eggs

Silk moths during oviposition lay both fertilized and unfertilized eggs due to various factors. It is very difficult to separate those eggs on egg cards. However, when the eggs are prepared in the loose form it can be separated by following a technique called brine/salt water treatment. Here the eggs are dipped in salt water solution having specific gravity of 1.06 and reject the floating eggs and then transfer these eggs to the salt solution having the specific gravity of 1.09 and collect the floating eggs (Fertilized egg will start to float).

Mother Moth Examination

Pebrine disease in *B. mori* is caused by a parasitic sporozoan, Nosema bombycis. Since pebrine disease is transovarially transmitted from one generation to the succeeding generation through mother moth. Hence elimination of pebrine infection, therefore, primarily depends on adoption, strict and careful mother moth examination. This can be achieved by following Fujiwara's method of mother moth examination at seed production centres.

Moth examination is of 2 types.

They are:

1. Individual Mother Moth Examination
2. Mass Moth Examination.

Individual Mother Moth Examination

Individual moth examination is ideal to check the disease but laborious. Thergore, this method is done during the preparation of reproductive seeds at seed multiplication centers. Individual moth is taken and crushed in a moth crushing set with 0.25 ml of 2 per cent potassium hydroxide solution. A drop of crushed fluid is taken on a glass slide and a cover glass is placed over the drop of fluid. Then, this is observed under a microscope with 450 - 600 x magnification. If the sample contains pebrine spore then such lot should not used for seed production.

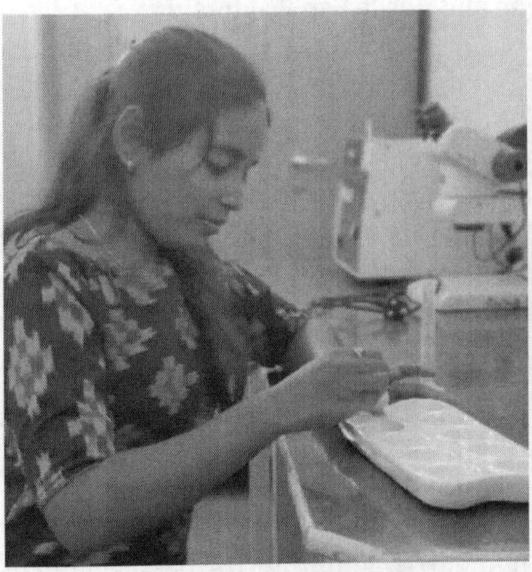

Crushing the Silk Moth

Filtering the Solution

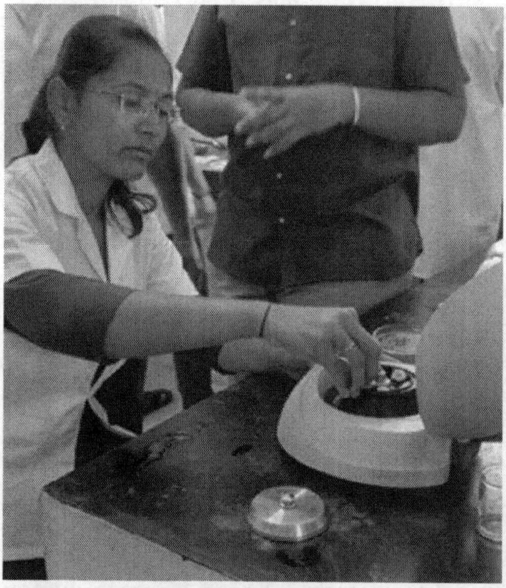

Centrifugation

Microscopic examination

The Mass Mother Moth Examination

Mass mother moth examination is quick and dependable method for pebrine disease and can be practiced in commercial grainages. In this case sample moths are drawn considering the number of moths in to a lot to be examined. Each sample comprises 20-30 moths.

Surface Sterilization of Eggs

After ascertaining the disease freeness of layings, egg sheets are dipped in 2 per cent formaldehyde for 10-15 minutes. This helps in removal of pathogens adhering to the egg shell and further prevents secondary contamination. It also aids in firm adherence of eggs to the sheets.

Preservation and Handling of Hibernated Eggs

In univoltine and bivoltine races, the eggs enter into diapuses in 40-50 hours after oviposition at 25 °C. Eggs when laid are yellow in color. Gradually they turn brown and then purplish brown. Such eggs do not hatch unless they are activated by cold temperature. This is naturally achieved in temperate areas. In tropical countries, the winter condition could be stimulated by preserving the eggs at required temperature in a cold storage. Such cold storage of eggs is carried out by following specific schedules. Low temperature preservation terminates the diapauses and activates the eggs to hatch.

The schedule comprises 3 distinct phases *viz.*, storing at 25 °C, cold storage at 5°C and intermediate phase, where the temperature is gradually lowered from 25°C to 5°C.

Short Term Refrigeration for 3 Months

The eggs are kept at 25°C for 3 days and then cold storage for 90 days at 5°C. The cold storage and release of eggs should be through an intermediate temperature of 15°C for 6-12 hours.

Refrigeration for 4 and 6 Months

The preservation period at different temperature is given below

Hibernation Period	*Temperature*							
	25°C	*20°C*	*15°C*	*10°C*	*5°C*	*2.5°C*	*5°C*	*15°C*
4 Months	10 days	2 days	2 days	3 days	50 days	50 days	-	1 day
6 Months	20 days	15 days	10 days	10 days	50 days	60 days	5 days	2-3 days

The eggs stored according to the above table can be released 15 days earlier or later than the scheduled period, without affecting the hatching. After releasing the eggs from 2.5°C, the eggs must be kept at an intermediate temperature of 15°C for one day and then transferred to incubation room.

Long Term Refrigeration (10 Months)

The schedule for preserving the eggs for 10 months is given in the following table

Hibernation Period	*Temperature*									
	25°C	*23°C*	*20°C*	*15°C*	*10°C*	*5°C*	*2.5 C*	*15°C*	*2.5°C*	*15°C*
10 Months	40 days	20 days	30 days	25 days	25 days	60 days	50 days	3 days	30 day	2-3 days

Acid Treatment for Bivoltine Eggs

Multivoltine breeds never undergo diapauses, but univoltine and bivoltines strains undergo diapauses or hibernation. Therefore to check the diapauses and to rear the worms in case of univoltine and bivoltine races, artificial hatching is done by two methods.

1. **Cold Treatment/Physical methods:** In this method eggs are allowed to complete the hibernation period under cold conditions.
2. **Chemical Method/Acid Treatment:** *i.e.,* cold or hot acid treatment Of these two methods, hot acid treatment is more popular.

Hot Acid Treatment

This treatment is practiced just between 15-20 hrs of oviposition *i.e.* before the egg show diapausing characteristics should be treated. When the treatment is to be postponed the eggs are preserved at 5°C at 20th hour. The period of cold storage should be within 5 days. During this period eggs can be taken at any time for acid treatment, but before the treatment eggs are exposed to 25-26°C for 2-3 hours.

Hydrochloric acid of specific gravity 1.075 (15 per cent) is heated to a temperature of 46.1° and the silkworm eggs are immersed in the acid for 5-6 min. The specific gravity is as measured at 15°C. Then wash the eggs in running water to remove traces of acid and dry under shade. "Now the eggs are ready for Incubation"

Treating the Bivoltine Eggs with Hot Hydrochloric acid.

12.10 Incubation

The purpose of egg incubation is to achieve uniform hatching on a desired day, by keeping the eggs at an optimum temperature of 25°C and RH of 75-80 per cent in a room already disinfected with 3 per cent formalin thoroughly or with 5 per cent bleaching powder.

Multivoltine eggs are incubated at farmers rearing houses by placing egg cards between two sheets of paraffin paper kept in a tray and a moist foam rubber strip is placed around the eggs when prevailing humidity is low.

Silkworm rearing includes activities from egg incubation upto harvesting and marketing of cocoons. Providing suitable environmental condition to the eggs for healthy growth and development of embryo is called as "incubation". Embryo grows robust using reserved food under 25°C and 80 per cent RH conditions. Hygienic condition such as surface disinfection of eggs ensures prevention from disease contamination.

Devices such as BOD incubators, earthern pots, double walled chambers or spreading of eggs on clean and disinfected paraffin paper in a tray bounded by wet foam pads are used for egg incubation.

Black Boxing

Light has profound role on growth of embryo and hatching. Providing total darkness for 48 to 60 hours before hatching is called black boxing. This helps in uniform hatching in a single day. During black boxing, those embryos which are in advanced stage of development wait for a light, to hatch and developing embryos will continue with their development, and when exposed to light, all the eggs hatch simultaneously.

The black boxing is done at the pin head stage (on 8th day of egg laying). The egg sheets are wrapped (25 or 50 dfls each) in a tissue paper and transferred to black boxes. The black boxes are kept in required humidity and temperature. The eggs are to be exposed to light, between 7 to 8 am on the expected day of hatching (10th or 11th day) to enable maximum uniform hatching.

Transportation of Eggs

- ✰ The transportation of eggs plays an important role in the development of embryo and successful crop harvest.
- ✰ The disease-free eggs are collected in wet handbags with proper aeration in early morning or in late evening.
- ✰ Transportation may be done during aestivation period, hibernation or post-hibernation period or pre incubation period.

IT Initiatives under Seed Sector

- ✰ Registration of Stakeholders under Central Seed Act: CSB has developed web based Online registration (new/renewal) process to facilitate the stakeholders *viz.*, Silkworm Seed Producers, Chawki Silkworm Rearers and the Silkworm Seed Cocoon Producers through www.csb.gov.in/ https://nssoregwebpages.firebaseapp.com, which eases the process of paperless submission/transaction for registration.
- ✰ "e Cocoon" mobile application: As a part of quick and real-time monitoring by the Seed Analysts/Seed Officers under Central Seed Act, CSB has developed an Android based mobile application "e-Cocoon "for onsite/ online reporting of the inspection proceeds of Seed Officers and Seed Analysts.

Chapter 13

Diseases of Silkworm/Silk Moth

Diseases of silkworm are a big challenge for the rearers. Due to prolong domestication, silkworm develops less immunity and less adaptability for which they are often attacked by different agents. Although some attempt has been made to develop disease resistant silkworm variety, but only a limited success is obtained till date. In spite of taking many efforts on the rearing environment, silkworms are vulnerable for infection of the following agents or pathogens

1. Protozoa
2. Bacteria
3. Viruses
4. Fungi

13.1 Protozoa

Causative Agent

A microsporidian, Nosema bombycis and strains of Nosema sp.

Phylum; Protozoa,Order;Microsporidia,Family; Nosematidae,Genus;*Nosema*

Species; *Nosema bombycis* Naegeli (1857)

Based on the microsporidian's unusual cytological and molecular characteristics such as primitive type of nuclear division and absence of mitochondria they have been phylogenetically considered as one of the earliest known eukaryotes. The historical developments on pebrine disease are arranged chronologically as follows:

Year	History
1750-1780	Extermination of sericulture industry in France due to outbreaks of pebrine
1763-1818	Boissier and Dandlo gave description of various diseases in which spots on the skin are cited as symptoms
1845	Existence of pebrine disease was known in France, Then its occurrence was observed in Italy, Spain, Turkey, Syria, Rumania *etc.*"
1849	Guerin - Meneville of France named the pebfine pathogen as "Hematozoid" ie, a microorganism living in haemolymph of silkworm
1850-1870	Leybig and Balbiani showed for the first time that Microsporidia are characterised by production of minute oval spores, In Europe the epidemic of pebrine swept the silkworm rearing districts
1856	Lebert and Fray made observations on pathological significance of the corpuscles
1857	Naegeli briefly destribed the ', corpuscles as representing some form of yeast -like fungus and gave the present name, Nosemabombycis
1858	Lebert studying the pathological significance of pebrine corpuscles, named the organism as Panhistophytonoyatam
1859	Hutton described the pebrinised, black spots appearing at the annulations of the body laying and also junction of the
1860	De Quadrefagues gave a popular name pebrine to this disease
1865-1870	Pasteur innovated transovarialtransmiSSion of pathogen and suggested mother moth examination for diagnosis of pebrine disease, He wrote a voluminous work entitled "ElUDES SUR, LA MALADIE DES VERS A SOIE", In this book thedisease has been called "MALADI CORPUS CULAIRE" (CORPUSCLE DISEASE)
1866	A disease in Mysore described as a sort of atrophy was also doubtlessly pebrine
1878	Pebrine epidemic swept through Kashmir
1882	Balbiani of France wrote a book "Lecan,'>SurlesSporozoaoirls" and made known that the pathogen belongs to Microsporidia under Sporozoa of Protozoa
1888	Mukerji stated that pebrineh~d appeared only 10, i, 12 years back and becoming more destructive in ead year
1890-1900	Epizootic outbreak of pebrine in the then M YS('l and Madras provinces
1899	Hutchinson wrote in all probability that pebrineh,b played a role in depressing the successful rearing of silkworm in India
1909	Stempel! from Germany published well know, paper describing the various stages in the life c)I' of*N. bombycis*
1913	Pailiot brought out that deadly epidemics v, caused by N. bombycis among silkworms in France

Year	***History***
1914	Basu reported that pebrine disease on erisilkw(''' Philosamiaricini was not much important
1916	Lefroy's Report on an enquiry into the silk industr ' in India stated this disease to have appeareGBengal by about 1895 and the probability in Mysore was between 1890 and 1900
1916-1920	Hutchinson (1918) opined that pebrine disease in India showed its appearance only towards the close of the last century. Infectivity, pathogenecity, transmission and viability of N. bombycis were studied
1917	Fauchere stated that silkworms in Europe and Madagascar were infected seriously by pebrine rendering more serious loss in twpics
1916	According to Deboudeau, pebrine appeared in Persia at about the same time as in Europe with the same disastrous results
1929	Flass working with tasar silkworm, Antheraeapernyi observed a disease, "Nosematose" caused by Nosema sp. closely resembling N. bombycis
1931	Condura working with one microsporidiosis of silkworm B. mori found that pebrine was responsible for serious losses in Italy
1955	Pebrine started as mildly epidemic in a large seed campaign during 1 94()"4 7 in Mysore
1960	Janakiraman stated that pebrine was not known until 1895 but he believed it must have existed during 1850-1870
1963	As per Yokoyama, pebrine was destructive to sericulture before Pasteur's principle was known and at present the actual damage caused by this disease is very slight in Japan
1992	Taxonomic position of *Nosema spp.* In the phylum microsporidia (sprauge *et al.,* 1992)
2010	Development of antibiotics for control with reference to diseases of silkworm.(Venkatesh Kumar *et al.,* 2010)
2013	A New methods using powder formulation of antibiotic and fungicide mixture to control pebrine
2013	Delayed mother moth examination methodfot detection of Pebrine,
2014	Molecular and Microscopic Detection of Microsporidia in some Silkworm (*Bombyx mori* L.) Populations in Turkey
2016	Quantitative PCR for detection of Nosemabombycis in single silkwormeggs and newly hatched larvae
2015	Detection of pebrine disease in *Bombyx mori* eggs with the loop-mediated isothermal amplification (LAMP) method based on *EB*1 gene in china
(Oct. 2016 - Sep. 2017)	Standardization and Validation of LAMP (Loop mediated isothermal amplification reaction) techniquefor the detection of *Nosemabombycis* infection in silkworm at CSRTT Mysore

Seasonal Occurrence

The disease is known to occur at any time of the year and is not restricted to any season.

Source of Infection

The faeces of the diseased silkworms and dead larvae. n Pathogen persisting in rearing environment, rearing house and appliances. n Contaminated mulberry and surface contaminated egg shell. n Mulberry and agricultural pests and wild insects as alternate hosts for the pathogen.

Mode of Transmission

The disease is transmitted to silkworm in three different ways

i) Oral- The leaves in the rearing bed normally get contaminated with the spores liberated through the faeces of infected worms or dead larvae. So, when such food is consume by healthy larvae, they get infected. Contamination may also occur at the time of oviposition or after oviposition.

Pathogenicity of different Microsporidia Infecting *Bombyx mori L.*

Microsporidian Isolate	*Pathogenicity*	*Spore Form*	*Spore Size (mm)*	
			Length	*Breadth*
N. bombycis	High	Oval	3.8	2.6
Nosema sp. M11	Low	Oval	3.9	1.9
Nosema sp. M12	Low	Ovo-cylindrical	4.5	2.0
Nosema sp. Lbms	Low	Ovo-cylindrical	4.36	2.14
NIK-2r	High	Oval	2.7	1.6
NIK-4m	Low	Oval	3.6	2.8
NIK-5hm	Low	Oval	5.0	2.1
NIK-5d	High	Ovo-cylindrical	5.0	3.1
NIK-1pr	High	Ovo-cylindrical	5.41	2.85
NIK-1Cc	Low	Ovo-cylindrical	4.60	2.77
NIK-1Cpy	High	Oval	4.60	2.85
NIK-1So	High	Ovo-cylindrical	4.96	2.61
NIK-1Dp	Low	Oval	5.26	2.79
NIAP-6p	-	Oval	4.27	2.40
NIAP-7g	-	Oval	5.00	2.50
Microsporidian sp. S1	High	Oval	4.60	1.01
Msp	-	Ovo- cylindrical	1.73	2.92

Major difference occurs between different strains of *Nosema sp*.

Strains	*No. of Coils*	*Tilth Angle*	*Structure*	*Raibosome Arrengments*
Nosema bombycis	12	49°	☆ Filament has central core surrounded by four concentric layers. The core is surrounded by a relatively electron lucent layer with a substructure cosisting of 16 smali granules. ☆ Spores, under light microscope are oval in shape, (approximately 3.4 x 1.5 - 2.5micron). With a three layered membrane (inner, middle and outer). ☆ The polarplast has two parts. At each end there is a vacuole. A polar capsule and polar filament are present. The polar filament is 30 times the length of spore.	The internal and the outermost layer are translucent like the membrane of endoplasmic reticuluor nucleus and between the two layers is an electron dense layer.
Nosema sp. M 11	11	39°	☆ Filament is composed of a core surrounded by three concentric layers. ☆ Spore has two nuclei placed close together. Nuclei are slightly elongate in the direction of the major axis of the spore. ☆ The polarplast consists of two parts as in the case *of N. bombycis.*	Ribosome arrangement is similar to that of *N. bombycis.*
Vairimorpha sp. (*Nosema* sp. M 12)	15	43°	☆ Structure of polarplast is similar to *Nosema*sp. M 11 but the polarplast is more slender than that of *N. bombycis*and *Nosema*M 11. ☆ The polarplast consists of two morphologically different components: an anterior part composed of laminated membranes, which may be tightly flattened sacs of loose shoe shape and a posterior part made up of loosely flattened sacs of lozenge shape. ☆ Spore possesses two nuclei in the posterior half that are slightly elongate in the direction of the minor axis of the spore.	An arrangement of ribosome in the cytoplasm just internal to the coiled polar filament
Pleistophora sp.	8	31°	☆ The filament has a central core surrounded by concentric layers. ☆ The polarplast is composed of three parts. The spore contains single nucleus in the posterior half. ☆ A small vacuole enclosed by many fold membranes occurs near the posterior pole and sometimes it has vesicles along its side.	Around the coil of polar filament long endoplasmic reticulum is often seen.

ii) Direct contact- Infection may occur through the skin of larvae in the rearing bed when contacted with the Nosema spores contained in the faecal matter or dead tissues.

iii) Transovarial- when infection occurs in the 5th instar larvae the adult moth immerges normally from the pupa. Under that situation the spores of Nosema Sporulate within the oocyte and pass on to the egg. Thus the eggs acquires infection from the mother and hatched into primary infected larvae. However, most of these infected larvae grown up to 3rd instar only. These larvae again act as the source of secondary infection through dead larvae and contamination by faecal matter. Secondarily infected larvae exhibit pebrine symptoms, unable to spin healthy cocoon and produce tertiary infection to 5th instar larvae.

Symptoms

Lack of uniformity in egg shape, reduction in egg size and weight, eggs being laid in clumps, poor egg attachment, high percentage of unfertilized, dead eggs, infection at the second and third instar and in the grownup larval stage, exhibited symptoms as wrinkled silkworms, *i.e.*, after feeding, larvae of the different instars show wrinkled skin with rusty colour; spotted silkworm ie., black pepper-like irregular spots appearing on the body mostly on the caudal horn and lateral to the prolegs are the symptoms. In moths diseased moths may develop clubbed wings, developed wings not spreading; moths with vesicles or black spots on wings, being dull, naked moths without abdominal scales, moths not showing mating instinct, discolouration of abdominal scales and shedding of scales from wings and body are also observed

Histopathologycal Changes due to Pebrine Infection

Malpighian tubules were more susceptible to the attack than any other tissue. In various tissues, the damage may be due to infection of muscles and nerves causing disturbance of their normal function rather than direct contact. When the pathogen invades the muscle, the tissues may be softened and liquified34. The parasite can also multiply within the layers of the integument. As the protozoan multiplies, the cytoplasm within the host cell may disappear slowly. The nucleus is rarely invaded. Severe infection of fat results in poor energy turnover and death of larvae or pupae may result. Infection of silkglands causes very poor synthesis of silk fibroin. Localization of pathogens in the muscular and nervous systems causes the disturbance to functioning of organs which leads to disorganization.

Management

- Two rounds of effective disinfection of rearing houseand appliances may have to be ensured once immediately after cocoon harvest 5 per cent bleaching powder solution.

- ☆ Timely elimination of lepidopterous defoliators of mulberry in the mulberry gardens may help to a great extent in checking the possible spread of pebrine pathogen from mulberry gardens to the rearing houses.
- ☆ Bed cleaning at regular intervals. Providing optimum spacing, temperature, ventilation and relative humidity to the worms during rearing.
- ☆ Timely picking of pebrinised worms and their destruction by dipping in 3 per cent formalin.
- ☆ Forced ec1osion may be achieved in grainage by exposing some seed cocoons of each lot to a temperature of 30 - 32° C which enables the moths to emerge 1 to 2 days earlier compared to the normally preserved seed cocoons.
- ☆ The early emerged moths are used for pebrine spore detection. This avoids further spread of the disease. Prompt burning of pebrinised layings, moth residues after mother moth examination and used male moths to prevent further contamination.

Mass Mother Moth Examination

The method of mass pebrine inspection of moths was adopted in Japan for the first time in 1968. In mass pebrine inspection 14-30 moths are crushed together and spores are separated from this mass. However, with this method, infection of a single moth among the mass results in the rejection of all eggs laid by the group of moths. Therefore, this method can be used only in the country where there is less possibility of the incidence of pebrine disease In mass pebrine inspection a large number of moths (20-30) are crushed together in a mixer with 80-90 ml of 0.5 per cent K2C03 solution. Then the moths are ground and centrifuged at 9000-10000 rpm for 2 min for spore liberation from tissues. The crushed material is allowed to stand for 2 min and filtered. The filtrate thus collected is centrifuged at 260 rpm for 3 min. The sediment is suspended in 2 ml of 2 per cent KOH solution if dry moths are used and in case of green moths a small quantity of 0.15 per cent KOH is added to the sediment which is put on a microscopic slide for microscopic examination (Geethabai *et al.,* 1985). Using automatic equipment called Hirano moth testing machine that has a grinding, a washing and pouring unit and a centrifuge, has been developed at Japan using which in a day of 8 hours about 500 samples can be tested.

An Improved Method for the Detection of Pebrine (*Nosemabombycis* N.) Spores in Silkworm (*Bombyx mori* L.) In Research Laboratory Method

The N. bombycis, collected from diseased silkworms, were propagated in Bombyx mori as per the method of Undeen (1997). Infected moths (20-30), in sterilized water in 0.5 per cent potassium carbonate (K2CO3) solution, were crushed in a mortar and pestle to separate spores from host tissues. The suspension was filtered twice through absorbent cotton and washed 2-3 times in endotoxin free (E.F.) water. Then suspension was centrifuged at 5000 rpm for 10 min. and sediments treated with 2 per cent potassium hydroxide (KOH) solution. The Nosema suspension was layered over percoll cushions and centrifuged at 12000-15000 rpm for 30 min. in an RC5C high speed refrigerated centrifuge (Sorvall) with swingout-rotor (SH-MT12) to obtain purified Nosema spores. Neubaurhemocytometer (German Fine Optik) with ThomaZaiss counting slide was used to count Nosema spores under compound microscope (15 x 40X) for determining the concentrations

Improved Mother Moth Examination Method

For the examination of mother moths, the standard method of Fujiwara (1984) was followed. The eggs laid by the apparently healthy pebrine-free mother moths were used for the production of nucleus (P1) and reproductive (P2) seed. Twenty moths were taken in a mixie cup and 80 mL of 0.6 per cent K2CO3 solution was added to it. Mixture was grinded and homogenized for 1-2 min. at 6000-8000 rpm. The homogenate was allowed to settle for 3- 5 min. in a beaker and un-macerated tissues and debris were separated and discarded by decantation. Bottom liquid was carefully filtered through thick layer of clean muslin cloth. The filtrate was re-centrifuged for 3 min. at 3000 rpm, the supernatant solution decanted and sediment dispersed in a few drops of 2 per cent KOH solution over a cyclomixer. Two smears from each sample were examined under phase contrast microscope (15 x 40X). The intensity of infection was estimated on the basis of the number of spores/field.

Delayed Mother Moth Examination Method

For delayed mother moth examination method (Sing *et al.,* 2004), the moths after oviposition were collected in groups in perforated cardboard boxes/covers and preserved in an isolated, well-ventilated room at 25-30°C for a period of 3-4 days and examined under microscope (15 x 40X) at 4 days interval. Moths were examined as per the procedure described above

New Methods Using Powder Formulation of Antibiotic and Fungicide Mixture

Live larvae/pupa/moths were taken in sterilized 5 mL 0.6 per cent K2CO3 solution and crushed in a mortar and pestle to separate spores from tissues. The suspension was filtered through absorbent cotton and followed by muslin cloth once when sample size was 1-2 larvae/pupae/moths and twice if the sample size was 3-10 larvae/pupae/moths. The suspension was kept at room temperature (25-33oC) for 30 min. when sample size was 1-2 larvae/pupae/moths and for 1 hr when sample size was >2 larvae/pupae/moth. The suspension was decanted when sample size was 2 larvae/pupae/moths. The sediment was mixed well with 5 mL 0.8 per cent KOH to dissolve fat body. Then sediment waskept at room temperature for 30-60 min. and refrigerated at 5±3°C for 6 hr. The chilled suspension was then washed twice by either decantation or centrifugation at 3800 rpm for 5 min. in E.F. water. The sediment was mixed well with 2 mL distilled water. A powder formulation of Ambistryn-S (broad spectrum antibiotic preparation of streptomycin sulphate) and Transco Trichostar (bio-fungicide preparation of Trichodermaviridae) (25 mg each in 5 mL preparation) was added to the suspension and shaken well or short spun by cyclomixer for 1 min. to mix the sample well. The suspension was kept at room temperature (25-30oC) for 15-30 min. and then in a refrigerator at 5±3°C for 3-24 hr to check the germination and growth of bacteria or fungi. The chilled suspension was then either decanted or centrifuged at 4800 rpm for 8-10 min. and washed in distilled water. The suspension was examined under microscope (600-1000X) and if any pebrine spore was noticed the whole sample was destroyed by dissolving the suspension in 5 per cent formalin or bleaching powder solution. The suspension may be stained with Nigrosin (10 per cent) for spore identification

A Field-Friendly Loop-Mediated Isothermal Amplification (FF-LAMP) Method

Pebrine is a destructive disease that exhibits horizontal and vertical transmission and therefore it is the only mandatory quarantine item in sericulture. Here, a field-friendly loop-mediated isothermal amplification (FF-LAMP) method has been developed and validated for the rapid detection of Nosema bombycis, a causative agent of pebrine disease in silkworm, Bombyx mori. FF-LAMP primers were selected and designed for small ribosomal subunit gene and the assay was performed to detect the N. bombycis infection in silkworm. The FF-LAMP reaction was effective at 6 mM MgSO4, 1.4 mM dNTPs at 63 °C. The detection range of LAMP assay was found to be 101 dilutions of N. bombycis spores. Specificity of the primers was tested using DNA isolated from pebrine infected silkworm, pebrine free silkworm and pure N. bombycis by conventional PCR and FF-LAMP assay. Results revealed that the primers were specific to N. bombycis DNA. The FF-LAMP assay was validated in different basic silkworm seed farms with simultaneous microscopic examination of N. bombycis infection. This newly developed method is highly effective, specific, sensitive and rapid in detecting N. bombycis infection, eliminating the DNA purification steps and usage of sophisticated equipment. This method can be used in testing large number of samples making it field friendly method in sericulture industry.

13.2 Bacterial Diseases of Silkworm

Silkworm rearing is a highly enterprising vocation for a large number of sericulturists across the country. The success of silkworm crop depends on many factors. The good quality mulberry leaf, proper temperature and humidity conditions, larval spacing, aeration, protection from diseases and pests, *etc.*, are some of the factors that contribute to the success of silkworm crops. Many silkworm diseases are caused due to the infection of various bacteria, virus, fungus, microsporidia, *etc.*, in the silkworm colonies and this is a major limiting factor in the successful cocoon production. Diseases such as grasserie, flacherie, muscardine and pebrine are caused due to these pathogens. Flacherie diseases are caused by microbial and amicrobial agents and microbial flacherie is caused by both bacteria and viruses. Bacteria are involved in causing flacherie as individual pathogen or in combination with viruses thereby contributing for disease epizootics. In China and Japan the silkworm crop loss due to diseases are less, but in India it is more and severely affects the economic returns of the rearers. Of the total disease occurrence in Karnataka the bacterial flacherie is to the extent of 57.22 per cent as assessed by microscopic examination of haemolymph and midgut smears (Samson *et al.,* 1990). Infection with several bacteria causes severe mortality and majority of the deaths occur during moult and when the worms are about to spin. Mortality becomes maximum among fifth instar silkworm larvae. Flacherie in silkworms is a syndrome characterized by flaccidity of larval body. The flaccidity caused by bacteria is referred as bacterial flacherie. Louis Pasteure reported bacteria as etiological agent of flacherie in silkworms as early as in 1870. The diseases in silkworms caused by bacteria are generally classified as bacteremia, septicemia and toxemia. For better appreciation, the topic is being explained into different sub-headings and they are:

Bacterial Diseases in Silkworms In 1870, Pasteur had reported that a coccus and bacillus bacteria caused the flacherie in silkworms. Later, many researchers have reported that different bacteria are causing bacteriosis in silkworms. Cuboni and Garibini (1890) thought that Bacillus cubonianus was the cause of flacherie. In Japan, the causal agent of Sotto disease was identified as Bacillus sotto (Iwashita, 1902), which presently is 1 known as *Bacillus thuringiensis* var. sotto. Sawamura in 1960 isolated *Bacillus megaterium* and Bacillus ellenbachi from the silkworms. Serratia marcescens (Metalnikov and Chorine, 1928), Bacillus bombysepticus (Hartman, 1931), Bacillus bombycoides (Paillot, 1942), Streptococcus bombycis (Steinhaus, 1949), Bacillus mucoides and Bacillus latrosporus (Steinhause, 1949), Bacillus cereus var. alesti (Tourmanoff and Vago, 1950), Aerobacter cloaceae and Achromabacter superficialis (Nakasuji and Kodama, 1969), Streptococcus faecalis, Serratia piscatorium, Proteus vulgaris, Proteus inconstina, Proteus morganii, Proteus aerobacter, Proteus aerogenes, Micrococcus flavus (Kodama and Nakasuji 1969; Nakasuji and Kodama, 1969; 1970; Vago, 1963) were also reported to be pathogenic to silkworm causing flacherie. Chitra *et al.* (1973) isolated Aerobacter cloacae, Achromobacter delmarvae and A. superficalis, Achromobacter delmarvae, Pseudomonas boreopolis, P. ovalis, Escherichia freundii and Staphylococcus albus. Flacherie in silkworms is a syndrome characterized by flaccidity of larval body. The flaccidity caused by bacteria is referred as bacterial flacherie. The symptom caused by different bacteria is atypical and are of general type. The larvae lose appetite, become lethargic and vomit gut juice. The growth becomes stunted and the larvae become flaccid. The dead larvae develop different colours, depending on the species of bacteria involved in causing the infection, become rotten and foul smelling. However, based on the site of infection and pathogenicity caused by the bacteria in silkworms, the bacterial diseases are categorized into three types. They are:

a) Bacterial Flacherie/Bacteremia

Bacteremia or bacterial disease of digestive tract is caused due to the multiplication of bacteria in the gut. The bacteria infect silkworm primarily through mouth and digestive tract and results in the development of a chronic type of flacherie. The cephalothoracic region becomes translucent and in the advanced stage even the posterior portion becomes translucent. The larvae show loss of appetite, sluggishness, retarded growth and develops diarrhea. Streptococcus faecalis and Streptococcus faecium are the most common bacteria associated with bacterial disease of digestive organ in the silkworm Bombyx mori. These are gram positive bacteria measuring 0.7 - 0.9 μm in diameter. The Streptococcus faecalis multiplies for 2-3 days and form large colonies attached to the larval peritrophic membrane. After 6 days, the peritrophic membrane dissolves and the goblet and cylindrical cells become vacuolated. The larva dies as the fluid in digestive tract and the bacteria passes through the vacuolated midgut epithelium into the hemolymph. In the digestive tract, the bacteria produce enzymes such as chitinase, lecithinase and proteinase which act on the midgut epithelium and enable the bacteria to enter hemocoel. Bacterial exotoxins and endotoxins also play a major role in invasion of the digestive tract by different bacteria. The toxins damage the gut wall and enable the bacteria to enter the hemocoel. Infection of S. faecium in fifth instar larvae result in lowering of gut pH from the range of 8.0 and 9.5 to 7.5 and 7.6. The lowering of gut pH is conducive for most septicemic bacteria.

Symptoms

i) Lack of eppetite. Diarrhoea and vomiting occur as common symptoms.
ii) Growth stunted and shrinkage of the body observed after moulting.
iii) Skin become loose and soft and exhibit sluggish movement.
iv) Body colour become dull to black-brown.
v) The larvae show signs of pain and convulsions and ultimately die.
vi) Body turns black and emits foul smell.

Mode of Transmission

i) Oral transmission- After taking contaminated leaves by the larvae from the rearing bed bacterial growth occur in the gut of larva and consequent deposition of lactate, acetate *etc.* reduces the gut permeability and lower the pH and infection spreads all over the body of the larvae. Larvae release stool with bacterial spores and then next set of infection starts.
ii) Rearing environment – Environmental physical factors like temperature and humidity of the rearing room also plays a vital role in disease transmission. Humid environment in the room may enhance the spreading of Flacherie bacteria.

Detection of Flacherie Bacteria

Histopathological analysis of gut can detect the bacterial infection by the detection of cylindrical or goblet cells in case of streptococcus and Diplococcus infection respectively.

Control Measures

i) Through maintaining optimum humidity and temperature in the rearing room.

ii) Providing healthy and disease free leaves to the rearing bed.

iii) Disinfection of the rearing room and the rearing appliances with 2 per cent formalin should be done regularly.

b) Septicemia

Septicemia in silkworms is caused by the septic bacterial invasion into the silkworm body and their multiplication, which results in the mortality of silkworms. The majority of silkworms that die from this disease vomit the gut fluid. The body of the diseased silkworm becomes soft, shrunken and the thoracic region swollen. A large number of bacteria are observed in the Haemolymph. The body wall ruptures easily liberating foul smelling fluid containing large number of bacterial cells. The larvae die in short span of infection. Several bacteria *viz.*, Serratia marcescens, Pseudomonas sp., Bacillus proteus, Bacillus aergenes, Bacillus prodigiousus, Bacillus pyocyanes, Streptococcus sp. have been isolated from silkworms affected with septicemia. But most common are Serratia marcescens and Bacillus sp. In case of septicemia caused by Bacillus sp., the silkworm shortly after death develop dark greenish tinge on the dorso-thoracic region which later extends to the whole body. In case of septicemia caused by Serratia marcescens, the larvae after death develop dark brown spots and reddish tinge, which extends to whole body. Injection of 6-8 x 102 bacteria causes death within 18 hours. The pathogenicity of orally inoculated Serratia causes a high mortality at high humidity, although the mortality varies with the amount of bacteria added to the feed. Bacterial involvement in septicemia of pre-pupal or pupal stages leads to the spoilage of cocoon quality and reduction in grainage productivity. The Bacillus species of bacteria causing septicemia are rod shaped ranging in size from 1-1.5 x 3 μm with sub-terminal spore and peritrichate flagella. Serratia marcescens appear as small rods measuring 0.6-1.0 x 0.5 μm. These are non sporulating bacteria with peritrichate flagella. Serratia marcescens in the digestive tract of silkworm are non-pathogenic. But when they enter the hemocoel they multiply rapidly and become pathogenic leading to acute disease and death in 1-3 days. Apart from the damaged gut wall the septic bacteria invade silkworms through wounds caused by injury and multiply rapidly in the hemolymph. The invasion by Serratia pisatorium into haemocoel in silkworm is observed to enhance by the presence of Streptococcus faecalis/Streptococcus faecium in the larval midgut. The lowering of pH by these bacteria favours the Serratia piscotorium to multiply and invade the hemocoel and produce fatal septicemia.

Symptoms

The larvae exhibit

i) Body exhibit softening.

ii) Body colour changes to brown.

iii) Larvae release liquid excreta.

iv) Loss of clasping power and death occurs frequently

Control Measure

Proper sanitation an hygiene should be maintain in the rearing room. Infected larvae should be immediately removed from the bed to prevent spreading of the disease. Handling of larvae should be done carefully to reduce skin wounds

c) Sotto

This disease is caused by some bioinsecticidal bacteria. For eg, Bacillus thuringiensis, which is widely use as a bioinsecticide release some endotoxin thuricide. When the larva consume the toxin two types of affects occur. In one type the larvae of B. mori stop feeding, haemolymph urn alkaline and papalysis occurs within 60-80 minutes of intact. In another type lepidopteran larva only show rapid inhibition of feeding.

Symptoms

Loss of appetite, sluggish movement, shrinkage of skin, diarrhea, Loss of clasping power of prologs, Body turns black after death and foul smell emit from the body.

Control Measure

Cleaning of rearing room with hot water to inactive the toxins, immediate removal of the infected or dead larvae for the rearing bed, bacterial spores should be destroyed by exposing to 2 per cent formaldehyde for 3hrs or 1000C for 5minutes

d) Court Disease

In this bacterial disease the affected dead larvae turn flaccid and the colour changes to crimson red. This disease is also commonly known as Rrangi due to this formation of colour.

Causative Bacteria

Disease is caused by the infection of bacteria Serratia marcescans and S. piscavotar following the primary infection with *Streptococcus faecalis.*

Infection occur through oral route or the skin wound

Control Measure

Maintenance of the hygienic rearing room, removal of dead or infected larvae, cleaning of appliances with 2 per cent formaldehyde are the measures for the diseases.

e) Bacterial Toxicosis

The cause of bacterial toxicosis is Bacillus thuringiensis. The common Bacillus involved is B.t. var. sotto. The larvae suffering from bacterial toxicosis will lose appetite and become sluggish. The dorsal vessel pulsates at rapid rate and the larvae wriggle as though they are in severe pain. Three exotoxins, *viz.*, α-exotoxin, β-exotoxin, γ-exotoxin and an endotoxin, δ-endotoxin have been isolated from B. thuringiensis. The poison ingested by the larva is first dissolved by the digestive fluid and partly absorbed by the body, which causes toxicity to the silkworms. This affects the nerves causing convulsions or paralysis. There are two types of bacterial toxicosis

i. **Acute bacterial toxicosis:** This is a poisoning caused by the ingestion of parasporal crystal bodies called digha-endotoxin produced by Bacillus thuringiensis. The major symptoms of bacterial toxicosis caused by Bacillus thuringiensis are – sudden cessation of feeding, lifting of heads, spasms and tremors, paralysis, distress, sudden collapse and death. Death may occur within 10 minutes to few hours. Shortly after the death, the corpse appears hard for touch with the head retracted so as to assume a hook-shaped appearance. Gradually the corpse become black and rotten, exuding a foul smelling dark brownish fluid.

ii. **Chronic bacterial toxicosis:** This is caused by ingestion of small quantity of Bt crystal toxin. Mulberry leaf intake is reduced, faeces become irregular shaped and occasional vomiting is observed and develops muscle paralysis. The thorax and abdominal tip become transparent and the worms become motionless. After 10-12 h of bacterial infection sluggishness is observed and loss of clasping power of the legs, flaccidity and death are observed

f) Mixed Infection of Bacteria with other Pathogens

Since the domesticated silkworms are susceptible to many infectious agents, mixed infections are very common. Mixed infection is the association of more than one pathogen causing diseases in silkworms. Synergistic interference of different pathogens influences the disease development in silkworms. The pathogen or its infection acts as a biological stressor and increase the susceptibility of the insect to another pathogen. The mixed infection play a major role in reducing the cocoon crop yields. Mixed infections may occur simultaneously or one after the other and usually the interaction between the two or more pathogens involved results in synergism thereby affecting the cocoon productivity both quantitatively and qualitatively. The silkworms very less susceptible to Serratia marcescens becomes highly susceptible when already being inoculated per orally with Staphylococcus sp. The reason being the lowering of gut pH by the Staphylococci thereby enhancing the multiplication rate of Serratia marcescenes. Similarly if the digestive tract of silkworms is attached with viral infection, the Bacillus bombycis otherwise non infective attains an infective proposition, causing damage to the silkworm crops. The symptoms of gattine are manifested only when the combination of Streptococcus and virus occurs in the susceptible silkworm.

'Thatte roga' a disease reported in Karnataka was found to be caused due to the synergistic association of bacteria such as Streptococcus faecalis var. liquifacians, Staphylococcus aureus, Staphylococcus epidermidis and Bacillus sp. and a picarnavirus namely Bombyx mori infectious flacherie virus (BmIFV). The disease was reported to begin in a localized area in the colony and spreads quickly at an alarming rate to the rest of the population. Due to this infection the larvae were observed to be feeble, lethargic and dull with translucent cephalothoracic region. They vomit gut juice and extrude soft faeces with high water content. The disease affects the later instar silkworms and the loss due to the disease was significant

Bacillus bombycis helps for the infection of nuclear polyhedrosis virus by making the condition of epithelial cells for virus infection and vice versa. The highest mortality of silkworms in simultaneous mixed infection is due to flacherie virus and bacteria. The susceptibility of BmIFV infected larvae increases in the presence of S. faecalis and Serratia marcescens. Similarly, simultaneous per oral infection of silkworm with BmDNV2 and Staphylococcus aureus results in higher mortality as well as deterioration in cocoon quantitative traits. It is well established that synergism between BmIFV and Streptococcus sp./Staphylococcus sp./Serratia sp. of bacteria and BmDNV and Streptococcus sp. bacteria results in the flacherie disease in silkworms. The synergism between different pathogens generally lowers the incubation period by 5-6 days and cause comparatively increased level of mortality. The bacteria such as Pseudomonas sp., Serratia sp. and Proteus sp. along with microsporidian spores in the larvae cause primary septicemia. The larvae generally die within 5 days day without developing microsporidiosis due to mixed infection.

13.3 Viral Diseases

a) Grasserie

It is one of the most dominant and serious viral diseases in tropical countries and occurs throughout the year. The *Bombyx mori* nucleopolyhedrovirus (*Bm*NPV) (Family: Baculoviridae; Genus: Alphabaculovirus) causing grasserie disease in silkworm is the most harmful virus in the sericulture industry, often causing severe economic losses [47, 7]. In India, >50 *per cent* of silk cocoon crop losses are attributed to *Bm*NPV infection [27] and in Kashmir valley the silkworm crop losses due to *Bm*NPV are about 28-32 *per cent* [19]. The nuclear polyhedrosis of silkworm is commonly known as Grasserie disease (French name), milky disease, hanging disease, fatty degeneration disease, or Jaundice (American name). In Italy it is known as "Giallume", in Japan "Nobyo" and in Germany "Gelbsucht". The names indicate either the yellowish colour of diseased insect or its swollen fat like appearance. In India, it is also known by various local names *i.e.* "Halu-hula" in Karnataka, "Rasa" in West Bengal, "Pola purugu" in Andhra Pradesh, "Paala poochi" in Tamilnadu and Aab-Baimair in Kashmir.

The grasserie disease of silkworm was first reported by Nysten. Earlier *Bm*NPV the causative agent of grasserie was referred as *Borriina bombycis*

Symptoms

This disease is characterized by the jaundice like symptoms which are collectively known as Grasserie- or Polyhedrosis. Other symptoms includes

- ☆ Swelling of inter segmental regions
- ☆ Larvae become lethargic and loose appetite
- ☆ Shining and yellowing body
- ☆ Hyperactivity
- ☆ Crawling around trays and hanging
- ☆ Rupturing of the skin
- ☆ White ooze filled with polyhedra
- ☆ Crawling around the periphery of the rearing tray
- ☆ Death of worms
- ☆ Diseased larvae loose the clasping power of abdominal legs, except the caudal legs by which it hangs with the head downwards.
- ☆ Rotting and secondary infections
- ☆ Nuclei of cells of various tissue contain polyhedral bodies
- ☆ The larva becomes restless and impatient causes fatigue

Causative Organism

The *Bombyx mori* nucleopolyhedrovirus (*Bm*NPV) is the most harmful virus in the sericulture industry, often causing severe economic losses. The nuclear polyhedrosis viruses belong to the sub group A of the family Baculoviridae. The family Baculoviridae is classified by the basic characters as an envelope, rod-shaped virion (approximately 50×250 micron) containing a circular double-stranded DNA genome ranging from 50 to 100 million Daltons. As the name implies, this virus multiplies and forms polyhedra in the nucleus of the tracheal epithelial cells, adipose tissue cells, dermal cells and blood cells. Occasionally the nucleus of the middle and posterior portion of silk gland cells are also affected. A characteristic feature of *Bm*NPV infection is the production of crystalline proteinaceous structures called occlusion bodies or polyhedra, in which several virions are embedded and thus protected from UV rays, desiccations, proteases and nucleases. Baculoviruses have only been found in over 600 species of arthropod hosts. The majority of baculovirus hosts are within the order Lepidoptera. They have also been isolated from the insects in orders Diptera, Hymenoptera, Coleoptera, Neuroptera, Thysanura and Trichoptera and the crustacean in order Decapoda (shrimp). The baculo portion of the name refers to the rod-shaped capsids of the virus particles. The virus particle is composed of a protein shell that surrounds the nucleic acid. Within the capsid, the DNA is condensed into a nucleoprotein structure known as the core. The capsid plus the core are collectively referred to as the nucleocapsid. Nucleocapsids contain a single molecule of circular super coiled double- stranded DNA. The length of baculoviral DNA is between 80 and 200 kb· Nucleocapsids are made in the nucleus of infected cells. The viral particles are rod shaped and the size is around 330×80 nm. The size of the polyhedra varies from 3-6 μ. The shape is usually octadecahedral or hexahedral and sometimes tetragon or trigon. Infection mostly takes place through feeding of polyhedra contaminated mulberry leaf, rarely through wounds. Factors inducing the outbreak of this disease are high temperature and humidity, their sudden fluctuations, bad ventilation in the rearing room, ineffective disinfection of rearing room and equipments and feeding of tender leaves during late instars. Inadequate larval spacing, starvation and excessive moisture in the rearing bed have also been known to contribute towards the outbreak and spread of the disease.

Mode of Transmission

i) Orally through contaminated leave with PIB from dead larvae or their excreta.

ii) High temperature and Humidity may enhance the transmission of latent stage of virus to virulent stage.

iii) Blocking of spiracle by some agents like dust, exposure to formalin *etc.* may enhance the spreading of the disease.

iv) Skin wound also enhance the disease.

Control Measure

i) Maintenance of hygienic environment, proper ventilation in the rearing room.

ii) Sterilization of the rearing room with 2 per cent formaldehyde or Bleaching powder or Resham Keed Ouzhad (RKO) or Labex.

iii) Immediate removal of dead or infected larvae from the rearing bed.

iv) Oral administration of nalidixic acid, P- aminobutyric acid *etc.* or topical application of imanine can control NPV to some extant.

b) Cytoplasmic Polyhedrosis Disease of Silkworm

The cytoplasmic polyhedrosis virus of silkworm *Bombyx mori* (*Bm*CPV) is a major viral pathogen of silkworm that causes extensive damage to the sericultural industry. Its incidence has been reported to be 27.76 *per cent* in Karnataka, out of the total loss of cocoon crop in India, 71 *per cent* was due to flacherie. The CPV disease is caused by an occluded double stranded RNA virus which multiplies in cytoplasm of columnar cells of midgut epithelium. The disease is characterized by the manifestation of symptoms like, body flaccidity, paleness of skin, vomiting and semisolid whitish green faeces. The severity of disease is mainly dependent upon the quality of mulberry leaf, silkworm breed and rearing conditions. The virus is

Bombyx mori Cytoplasmic Polyhedrosis virus (*Bm*CPV), which belong to the CPV subfamily, which consists of 19 distinct species (electropherotyes) within the genus Cypovirus, family Reoviridae· The genome of *Bm*CPV is composed of 10 discrete double stranded RNA segments. The virus particle is globular, 60-70 nm in size, icosahedral, hexahedron, each capscmer extends outwards to form a four-joined projection, the tip of which carries a globular body concealing the two distal joints. The capsid consists of two icosahedral coats which are linked together by a tubular structure at the supposing apices. At the centre of the capsid lies the core. *Bm*CPV infects epithelial and columnar cells of the midgut of susceptible silkworm. The infected silkworms are characterized by hypogenesis, emaciation and sluggishness. As the infection advances, white wrinkles can be observed in the posterior part of the midgut, which is the typical symptom of CPV caused disease

Mode of Transmission

Infection is mainly through oral route, the virus and polyhedra enter the digestive tract with the mulberry leaves that are swallowed. The formation of polyhedral proteins also takes place in the cytoplasm, after which the virions are enclosed to form polyhedra, which line up in rows in the cytoplasm.

After gaining entrance into the body, the *Bm*CPV first parasitizes the tubular cells at the junction of the mid and hind gut, causing a milk-white fold to form at this site. As the disease develops, the milk-white portion advances rostrally until the entire midgut becomes a milk-white annular prominence, giving the body of the worm a milk white colour. After infection of the silkworm by the *Bm*CPV, the digestive and absorptive functions of the midgut are impaired, which together with the expenditure of large quantities of host proteins to form viral and polyhedral proteins, cause disturbances in the nucleic acid and protein metabolism. As a result, the free amino-acid content of the haemolymph, the digestive fluids, the tissues and the cells decrease, and there even occurs an absence of some amino acids. All this leads to physiological disfunction and disease. Progression of the disease sees the tubular cells becoming packed with polyhedra, causing the cells to swell and finally burst. The polyhedra and viral particles released into the cavity of the intestine are excreted in the faeces, which assume a milk white colour.

Symptomatology of Cytoplasmic Polyhedrosis

Slow development of symptoms and extended courses are characteristics of the disease. Small worms exposed to minute quantities of the virus may not show sickness until the fifth instar. Generally, if infection occur at the first instar, the onset of the disease would be seen in the second and third instar, if at the second instar, than the onset would be in the third and fourth instar, if at the third or fourth instar, then symptoms appear in the fifth instar and it at the fifth instar, no symptoms are seen at all.

- ☆ Larvae show symptoms of sluggishness, loss appetite, cease feeding and sometimes develops diarrhoea and vomit gut juice.

- ✰ Larvae lag behind in their growth and development.
- ✰ Head becomes disproportionately large and translucent and body shrunken.
- ✰ Frequent expulsion of semi-solid and whitish feacal matter.
- ✰ In infected larvae midgut becomes chalky white and opaque.
- ✰ Infected pupae are generally smaller than normal and the infected moth lays fewer eggs and as short lived.
- ✰ On dissection of infected larvae midgut is seen as whitish and opaque compared to greenish and transparent midgut of healthy larvae.
- ✰ Anal region is solid and occasionally rectal protrusion is noticed.
- ✰ If infection taken place at the adult larva stage, the thorax becomes slightly transparent, the body atrophies, and vomiting and diarrhoea occur.

The course of the disease is influenced by temperature, humidity, and quantity and virulence of the virus. In the fields, exposure usually occurs at the third and fourth instar, and the onset of disease at the fifth instar. This disease occurs mainly during the summer and autumn rearing seasons.

c) Infectious Flascerie-Infecious Flasceri Virus (IFV)

Flacherie (literally: "flaccidness") is a disease of silkworms, caused by silkworms eating infected or contaminated mulberry leaves. Louis Pasteur, who began his studies on silkworm diseases in 1865, was the first one able to recognize that mortality due to viral flacherie was caused by infection. Bacteria is an etiological agent of silkworm flacherie. The flacherie in silkworms is by virus or by bacteria alone, or it can also cause by the combined infection of viruses and bacteria.

The major diseases affecting silkworm are flacherie, grasserie, muscardine and pebrine. Among these four diseases, flacherie is more prevalent causing cocoon crop loss to the tune of 33.88 *per cent* in India. *Bombyx mori* Infectious flacherie virus (*Bm*IFV) is a single stranded RNA virus that causes flacherie-like symptoms, is the most frequent virus in *Bombyx mori*, and is known to cause crop losses of up to 40 *per cent*[1]. The infected larvae show gradual reduction in size, retarded growth, reduction in body weight and flaccid condition of body followed by vomiting and death. Adverse environmental conditions such as high temperature and humidity, polluted air and starvation are considered important predisposing factors for flacherie and cocoon crop loss. Temperature higher or lower than 25 °C acts as a stress factor and increases the susceptibility of silkworm to viral infections. If temperature and humidity are extremely high, the susceptibility of silkworm larvae increases to viral infections

Causative Agent

This disease is caused by the *Bombyx mori* Infectious Flacherie Virus (*Bm*IFV), belonging to the family Picornaviridae. The virus particle is globular, about 30nm in diameter. The nucleic acid is single stranded RNA, no polyhedra are found in the host's cell.

Mode of Transmission

Infection is mainly by the oral route. After invasion of the midgut, the virus first infects the goblet cells, starting at the rostrum end and spreading upwards, without forming polyhedra. *Bm*IFV chiefly infects the goblet cells of the midgut, which perform the function of secreting digestive fluids, which in turn not only digest mulberry leaves but also have bacteriostatic and virucidal properties. Degeneration of the goblet cells affects the digestive and bacteriocidal functions and this permits rapid proliferation of enteric bacteria. Under the concerted invasion of virus and bacteria, death of the silkworm is accelerated. After infection, the goblet cell swells and later atrophy. The cells eventually disintegrate; the virus is then dispersed in the lumen of the digestive tract and is excreted with the faeces. Thus, the faeces, which carry large quantities of the virus, are an important source of infection within the rearing tray.

Symptomatology of Infectious Flacherie Disease

The symptoms and course of the disease vary, depending on the kind and quantity of concurrent bacterial proliferation in the digestive tract. The course of the disease is uncomplicated infectious flacherie is relatively protracted, lasting about 5-12 days. The symptoms of this disease and those of cytoplasmic polyhedrosis are rather similar.

- ☆ At the early stage of infection, symptoms are not clear and difficult to identify.
- ☆ At the initial stages of the disease consumption of mulberry leaves declines, the time needed to finish moulting is not uniform, and individual disparity in size is great.
- ☆ The larvae become soft and flaccid, the growth of affected larvae retards, become inactive and vomit gut juice.
- ☆ The faeces become soft with high moisture content. Sometimes chain type of excreta is observed. Often, rectal protrusion is also observed.
- ☆ Cephalothoracic region becomes translucent.
- ☆ The main symptoms are shrinkage, empty foregut, diarrhoea and vomiting.

  When infected with *Bacillus thuringiensis*, symptoms of toxicity such as paralysis and sudden death are observed. After death larvae turn black in colour and give foul smell.
- ☆ Sometimes the dead larvae turn red when infected with *Serratia marcescence* during injury.

The disease can be control by the following measures

- ☆ Feeding the larvae with healthy and hygienic leaves.
- ☆ Maintanence of proper hygiene of the rearing room.
- ☆ Disinfection of rearing appliances with formalin.

d) Densonucleosis Viral Disease of Silkworm

Viral diseases of silkworm cause a major problem to sericulture as they account for almost 70 *per cent* of total crop loss. The two major viral diseases include grasserie caused by *Bombyx mori* Nuclear polyhedrosis virus (*Bm*NPV) causes 42.60 *per cent* crop loss and viral flacherie caused by *Bombyx mori* Infectious flacherie virus (*Bm*IFV) causes 47.90 *per cent* crop loss, *Bombyx mori* Densonucleosis virus type 1 (*Bm*DNV-1) and *Bombyx mori* Densonucleosis virus type 2 (*Bm*DNV-2) causes 9.50 *per cent* crop loss [34].

Densoviruses (DVs) belong to the family Parvoviridae. Viruses belonging to the Parvoviridae family are known for having a wide host range. Hence this family has been divided into two subfamilies namely, Densovirinae and Parvovirinae. Viruses falling under the category of Densovirinae infect invertebrates while those under Parvovirinae infect vertebrates. *Bm*DNV-2 is a non-enveloped spherical DNA virus 20-24 nm in diameter and the genome contains two non- homologous single-stranded linear DNA molecules VD1 (6,543 bp) and VD2 (6,022 bp), which are encapsulated into separate virions.

In managing viral diseases, it is difficult to identify the pathogens causing the diseases and symptoms at early stages of the disease. Modern biotechnological tools like molecular and serological diagnosis for early detection of the virus in at early stages of silkworm growth play a critical role in management of virus.

Viral diseases of silkworm cause a major problem to sericulture industry. Grasserie and viral flacherie are the two major viral diseases. Grasserie is caused by *Bm*NPV and viral flacherie is caused by *Bm*IFV, *Bm*DNV, *Bm*KV or *Bm*DNV-2. Kenchu is a type of viral flacherie prevalent in Karnataka, India. The characterization of *Bm*DNV-2 revealed that it was a single stranded (ss) DNA virus and its bipartite genome consists of two linear ssDNA segments, which are packed into separate capsids.

Causative Agent

Kenchu virus is a small, non-enveloped, linear single stranded DNA virus, which was earlier classified as *Bombyx mori* Densovirus type 2 (*Bm*DNV-2) of the subfamily called Densovirinae which comes under Parvoviridae family. The genome of *Bm*DNA-2 consists of two DNA segments of VD1 (6.5 kb) and VD2 (6 kb). Unlike any other known ssDNA virus, instead of the typical rolling circle replication initiation endonucleases (RCRE), *Bm*DNV-2 encodes a type B DNA polymerase and has been suggested to replicate its genome via the protein-primed mechanism and is the only virus that carries a single stranded DNA that encodes a DNA polymerase. According to these unique features, in 2012, the International Committee of Taxonomy of Viruses (ICTV) voted to agree with the proposal to create a new genus, Bidensovirus, as well as a new family, Bidnaviridae, and designated *Bombyx mori* Densonucleosis-2 (*Bm*DNV-2) as *Bombyx mori* Bidensovirus (*Bm*BDV), the type species of the Bidensovirus genus.

Infection and Transmission

Bombyx mori Densonucleosis-2 (*Bm*DNV-2) infect the columnar cells of midgut epithelium and cause the fatal flacherie disease and replicates predominantly in the nuclei of columnar cells of the larval midgut epithelium. Evidence showed that a gene nsd-2 encoding a 12-pass transmembrane protein only expressed in midgut was responsible for the binding of *Bm*DNV-2 to silkworm midgut. In diseased larva, color of the postero-midgut changed to pale yellow. The main lesions occurred in the nuclei of infected cells. At the late stage of infection, the nucleus of infected cell was 2.5 times larger than that of normal cell and a voluminous dense homogenous structure appeared in each of the infected nuclei. At the late stage of infection, infected columnar cells degenerated and were discharged into gut lumen, and then the mature virions in the degenerate infected cells were egested along with faeces.

The ultra structural changes in cells infected by *Bm*DNV-2 were observed both in the cytoplasm and the nucleus. In the cytoplasm, accompanying a number of free ribosomes, vesicles or cisternae appeared near the granular endoplasmic reticulum, and small spherical particles accumulated inside of these microbody like structure. This step could represent the accumulation and transportation of viral protein to nucleus. Following the multiplication of virus, cellular organelle such as mitochondrion and endoplasmic reticulum became vacuolization and degeneration. Mitochondrion became hypertrophic and lost its cristae.

At the end of infection, the greatly enlarged nuclei were filled with thousands of small spherical virions with 20-24 nm in diameter, replacing completely the whole nuclear materials. The virions entered the cytoplasm through the enlarged nucleopores or by destruction of the nuclear membrane. Then, the complete virions were liberated into the gut lumen along with the degenerated cells.

Symptomatology of Densonucleosis Disease

The symptoms of this chronic disease affecting the early instars are such that farmers tend to continue rearing diseased worms for long periods of time without favourable returns as the worms will have to be discarded, this disease causes a loss of 9.50 *per cent* out of a total crop loss of 20-40 *per cent* in India. The recent discovery of resistant gene nsd-2 in some of the *Bombyx mori* races has led to the initiation of screening of resistant parental races for the development of *Bm*DNV-2 resistant transgenic breeds and these breeds were used as a parental breed in breeding programmes for developing *Bm*DNV-2 resistant silkworm races. The symptoms of *Bm*DNV-2 infection look apparently normal just after infection but slightly dull and redness of skin will be observed after two hours. Sometimes thoracic region looks swollen. In later stage of infection, vomiting gut juice and faeces with high moisture content is noticed. Early in stars are more susceptible to the disease and among different levels. Epizootiological and serological studies have found that the virus persist in the soil, gut of chicks and rearing trays.

Management of Viral Diseases of Silkworm

Viral diseases of silkworm are very difficult to manage due to a very short life cycle of silkworm. Several prophylactic measures are recommended to prevent the spread of the disease in the silkworm colony by the application of bed disinfectants such as Resham Keet Oushad (RKO), Vijeta, and Ankush *etc.* Although the disinfection of silkworm rearing environment and application of bed disinfectant is expected to maintain the pathogen load to the significant tolerable level, it may not be always adequate to prevent the diseases occurrence. Prevention is better than cure is the correct approach that should be adopted in silkworm disease management. Care is also needed to be taken to see that the silkworms are not exposed to stress conditions like temperature, humidity, bad ventilation, feeding arrest, parasitism and nutritional deficiency which may make them easily susceptible to grasserie disease [29]. Disinfection before, during and after each rearing should be carried out effectively and strictly.

For effective prevention of this disease, the silkworm rearing rooms, mulberry storage rooms, mounting rooms, equipments and rearing premises should be thoroughly disinfected before brushing. The eggs should be essentially surface disinfected. Silkworms should be reared under strict hygienic conditions. During rearing the diseased and dead larvae form the major source of infection with the largest quantity of fresh polyhedra available. Hence, the diseased larvae should be removed carefully without breaking the skin and disposed suitably by putting them in lime vats or by burning. Depending upon the suitable temperature and humidity should be provided. Fresh air circulation should be ensured by providing cross ventilation. The silkworms should be fed with nutritive rich mulberry leaf and during later stages feeding of tender leaf should be avoided. Do not provide over matured/over stored/dirty leaf to the silkworms. Avoid starvation, overcrowding and accumulation of faeces in the rearing bed. Depending upon the stage of larvae, spacing and required quantum of leaf should be given.

e) Gattine

This disease is also known as Clear head disease as the affected worm shows transparent, particularly the anterior part of the body. Causative organism is primarily some virus while Streptococcus bombycis act as the secondary agent. Loss of appetite and vomiting of alkaline clear fluid are the other symptoms. Supplying of proper healthy leaves, hygiene in the rearing room and Disinfection of rearing appliances are the measures to be taken for the controlling of the disease.

Some Useful Characteristics of Viral Diseases of Silkworm

Disease	*Causative Agent*	*General Symptoms*	*Some Distinct Characteristics*		
			Haemolymph	*Midgut*	*Integument*
Nuclear polyhedrosis	*Bombyx mori* nuclear polyhedrosis virus (*Bm*NPV)	Swelling on inter-segmental region; shining and fragile skin; milky white fluid; cadaver's grayish black and flaccid.	Turbid and white, occlusion bodies are present.	Not distinct	Shiny just before moulting; fragile; exuding fluids; in late instars, spiracles black and cuticlewith black spots.
Cytoplasmic polyhedrosis	*Bombyx mori* cytoplasmic polyhedrosis virus (*Bm*CPV)	Translucent cephalothoraxic region; diarrhoea; retarded growth; milky white midgut; whitish faeces; larvae pale, thin and small; growth slow and asynchronous; faecesremain attached to the anus; cadavers flaccid.	Clear, no occlusion bodies present.	Posterior midgut ivory in colour, shrunken, and striated, occlusion bodies are present.	Not distinct
Infectious flacherie	*Bombyx mori* infectious flacherie virus (*Bm*IFV)	Larvae vomit; head raised; growth and moulting asynchronous; larvae discoloured, flaccid; faeces watery; cadavers have a flattened shape.	Clear, without occlusion bodies.	Contains yellow-brown fluids; pink viral spherical bodies present; stains with pyroxinmethyl green.	Not fragile.
Densonucleosis	*Bombyx mori* densonucleosis virus (*Bm*DNV)	Thorax mostly translucent; like flacherie or bacterial flacherie anorexia; late moulting; bodies shrunk; integument not fragile.	Clear, without occlusion bodies.	Contains yellowish green fluids; cocci and diplococcic very abundant but noocclusion bodies.	Not distinct.

13.4 Fungal Diseases

a) Muscardine

In muscardine the body of the larvae get mummified due to deposition of calcium oxalate. Hence the disease is also called as 'calcino'. Symptoms includes- Sluggish movement of the larvae, loss of appetite, vomiting, shrinkage of body with loose cuticle are observed. Larvae infected at the early stage do not spin, but those infected late may spin cocoon where the pupa die inside. Dead larvae generally covered with the spores of the infected fungi or may be mummified with white powdery material. Causative fungi is different muscardine. Depending on the colour of the conidiospores name of muscardine is different.

Name of muscardine causative fungi

1. White muscardine - *Beauveria bassiana,*
2. Black muscardine - *Metarrhizium arisopliae.*
3. Yellow muscardine - *Poecilomyces farinosus*
4. Brown muscardine - *Aspergillus flavus and A. oryzel*
5. Red muscardine - *Sorosporella uvella*

Mode of transmission- all types of muscardine infection occurs through contaminated leaves, through skin and spiracles. The fungal spores germinate after falling on the silkworm skin and gradually penetrate the cuticle by mechanical and enzymatic forces

Control Measures

i) Proper light, humidity and aeration should be maintained in the rearing room. Any factors that may help the growth of fungi should be avoided. Temperature should be kept above 220C.

ii) Infected larvae should be remove immediately and buried in a deep pits.

iii) For the preparation of bed 0.45 formalinised husk can be use for 1st and 2nd instars, 0.5 per cent for 2nd and 3rd instars. 0.6 per cent for 4th instars and 0.8 per cent for for 5th instars.

iv) During moulting after every cleaning operation dithane-M45and kaoline can be applied to prevent the germination of fungal spores if any.

v) For every 100layings, 3-4kg of Resham Keed Ouzhad (RKO) can be spread after each moulting and 30 minutes before the larvae resume feeding.

vi) Laabex, a mixture of lime and bleaching powder has anti-muscardine and grasserie effects, also improve larval growth.

b) Aspergillosis

Aspergillosis is a fungal disease in silkworm which occurs during Chawki stage (I and II instar worms).

Causative agent : *Aspergillus flavus and A. tamarii*

Route of infection : Skin, the conidia from infected silkworm/alternate hosts contaminate rearing environment, rearing house and rearing bed. Infection starts when conidia comes in contact with silkworm body

Predisposing factors : High humidity and high temperature Symptoms : Infected larvae stop feeding, become lethargic and the larvae die soon due to aflatoxin produced by the fungus in the host. Aerial hyphae appear a day after death and later conidia cover the body giving particular colour according to the Aspergillus species (dark green in case of A. flavus and dark brown in case of A. tamarii)

General measures for the prevention and management of silkworm diseases :

- ✰ Practice two times thorough disinfection of rearing house, its surroundings and appliances with any recommended disinfectant (Asthra/Sanitech/Serichlor/Serifit/Decol) Select carefully the disease free layings
- ✰ Maintain personal and rearing hygiene during silkworm rearing
- ✰ Collect the diseased larvae at early stage of infection and ensure their proper disposal
- ✰ Maintain optimum temperature and humidity in the rearing house
- ✰ Feed good quality mulberry leaf, avoid overcrowding in the rearing bed and provide properv ventilation in the rearing house Apply bed disinfectants as per schedule and quantity (Ankush/Vijetha/Rakshak/Resham Jyothi) Feed Amruth (botanical based curative formulation against silkworm diseases) as per schedule and quantity

Chapter 14

Silk Reeling

14.1 Introduction

In general silk reeling is defined as unwinding of silk cocoon. However, it is technically defined as the process of finding the right end of the cocoon filament and jointly taking several ends together to reel raw silk. These processes are carried using reeling machines.

Depending on the size of the yarn required to be produced, filament from the cocoons are drawn out simultaneously, compacted and wound on to a reel. For example, if 20/22 denier yarn is required, filaments from 8 to 9 cocoons are drawn out simultaneously, compacted and wound.

Silkworms are reared for obtaining cocoons which form the raw material for producing raw silk. The technique of raw silk making was discovered several million years ago. The cocoons are softened using hot water and the silk filament is drawn continuously. This uniform thick and strong thread is used for the manufacturing of fine silk fabrics. Silk reelingis the final and purely commercial phase of sericulture. It is concerned with unwinding of the silk filaments of the cocoon.

The term raw silk (**Grege**) in industry or commerce is commonly understood to denote mulberry raw silk. It is the compact untwisted and undegummed silk thread that is formed by combining the required number of silk filaments drawn from as many separate cocoons by a special tech-nique called Reeling.

This includes a series of skilled operations to trans-form raw material (cocoons) into a fine, continuous silk filament of great length. Unlike other aspects of sericulture, reeling needs constant attention and care, since the process of reeling the cocoon filaments break continually and reeler must properly attach fresh filaments to make a continuous filament. If not, the reeled silk results in the abrupt occurrence of thin lengths.

The building in which cocoons are reeled for the production of raw silk is called a filature. It is carried with sophisticated automatic machines,to ensure production of raw silk of desired qualities. The filature concept isseen in developed countries where the raw material (cocoons) is of superior quality. But in other countries (where cocoon quality is not good) highly efficient multiend reeling machines with auxiliary equipment are installed and rational techniques of production are adopted. Besides these cottage industries are also found in the remote rural areas of cocoon production.These machines consist simple mechanism, simple methods of reeling.

The appliances are either foor-powered or hand-driven, with simple methods of raw silk production. The cottage reeling industry has neither set patterns of organization not standardized equipment and production techniques. But the aim of reeling in the cottage sector is always to obtain the maximum yield of raw silk from cocoons at the minimum cost of material and labour. It also produces raw silk which would satisfy all classes of silk weavers and other pure silk industries.

The new technology of handling silkworms in the country has shown that bivoltine silkworms, producing the international quality of silk, can be produced throughout the year in the Karnataka state in India. It has been shown that, on an average, 30 to 35 kg of cocoons, yielding 3 to 4 kg of high-grade silk can be produced by rearing 40,000 eggs of bivoltines, as compared with 25 kg of cocoons and 1.4 kg of low-quantity fibre from multivoltines.

14.2 Composition of Silk Cocoon

Silkworm cocoons are composed of proteins, which account for more than 95 per cent of its content; other impurities, such as waxes, mineral salts, and ash, constitute about 4-5 per cent. The wild silk cocoon shell has a lower sericin content and higher levels of wax, minerals wax, ash and other impurities. The raw silk fibre extracted from silk cocoon is subjected to degumming process to remove sericin from it. After degumming silk fibre contains fibroin protein

14.3 Composition of Silk

The silk is almost a pure protein composed of two types of proteins *viz.*, fibroin (70-80 per cent) and sericin (20-30 per cent) and other components such as carbohydrates, waxes and ash (1-2 per cent). Sericin (C15H25N5O8) comprises of 18 amino acids, with strong polar side groups and high content of serine, aspartic acid and glycine resulting in hydrophilic protein. The silk fibre after degumming contains fibroin protein which is composed of about 20 amino acids. Fibroin ($C_{15}H_{23}N_5O_6$) is made up of two components *viz.*, crystalline and amorphous. In crystalline component, amino acids are present in a definite manner with a definite space and the glycine residues repeats with other amino acids in the ratio of 3:2:1.

Japanese scientists have revealed that fibroin is composed of major homogenous part called heavy chain or H- chain (350KD) and minor heterogenous part called light chain or L-chain (25 KD) connected to each other by disulphide linkages. In mulberry silk, glycine, alanine, and serine together constitute about 82 per cent, of which serine constitute 10 per cent. Tyrosine and valine may be considered next to these at about 5.5 and 2.5 per cent, respectively. The dominance of acidic amino groups (*i.e.*, aspartic and glutamic acids) in the mulberry variety is more than that of the basic amino acids. The ratio of basic to acidic amino acids is 0.65 for the bivoltine and 0.75 for the crossbreed variety.

Amino Acid Composition of different Varieties of Silk Fibres

Amino Acid Composition (mol per cent)					
Amino Acid	*Mulberry (Bivoltine)*	*Mulberry (Cross Breed)*	*Tasar*	*Muga*	*Eri*
Aspartic acid	1.64	1.49	6.12	4.97	3.89
Glutamic acid	1.77	1.53	1.27	1.36	1.31
Serine	10.38	10.85	9.87	9.11	8.89
Glycine	43.45	43.73	27.65	28.41	29.35
Histidine	0.13	0.15	0.78	0.72	0.75
Arginine	1.13	1.16	4.99	4.72	4.12
Threonine	0.92	0.76	0.26	0.21	0.18
Alanine	27.56	28.36	34.12	34.72	36.33
Proline	0.79	0.76	2,21	2.18	2.07
Tyrosine	5.58	5.76	6.82	5.12	5.84
Valine	2.37	2.89	1.72	1.5	1.32
Methionine	0.19	0.11	0.28	0.32	0.34
Cystine	0.13	0.12	0.15	0.12	0.11
Isoleucine	0.75	0.78	0.61	0.51	0.45
Leucine	0.73	0.75	0.78	0.71	0.69
Phenylalanine	0.14	0.18	0.34	0.28	0.23
Tryptophan	0.73	0.75	1.26	2.18	1.68
Lysine	0.23	0.25	0.17	0.24	0.23

14.4 Physical Properties of Silk

Colour

Mulberry silk fibre is white or yellow or light green. Muga silk is golden yellow, tropical tasar is copperish yellow or fawn coloured, oak tasar is yellow grey and eri silk is white, dull white. White or colourless fibres are more preferred because they can be dyed or printed with any hue of colour.

Shape

Cocoon shape, as colour, is peculiar to the given species. At the same time shape can be affected by the execution of the mounting process, especially during the cocoon spinning stage. Generally, the Japanese species is peanut-shaped, the Chinese elliptical, European a longer elliptical and the polyvoltine species spindle-like in appearance. Hybrid cocoons assume a shape midway between the parents for example, the case of a longer ellipsoid or shallowly enclosed peanut form. The shape of cocoons assists in identifying the variety of species plus evaluating reelability

Wrinkle

The deflossed cocoon has many wrinkles on its surface. Wrinkles are coarser on the outer layer than within the interior layer. The outline of the wrinkle is not uniform, but various according to species and breeding conditions. Spinning employs high temperature and low humidity settings, which render fine wrinkles or cotton-like textures of, cocoon layers. These provisions discourage the agglutination of the baves resulting from accelerated drying. It is recognized that coarse wrinkled cocoons reel poorly.

Size

Cocoon size or volume is a critical characteristic when evaluating raw materials. The size of the cocoon differs according to silkworm variety, rearing season and harvesting conditions. The number of cocoons per litre, ranging between 60 and 100 in bivoltine species calculates size. Multivoltine species measure considerably higher.

Cocoon Weight

The most significant commercial feature of cocoons is weight. Cocoons are sold in the marketplace based on weight as this index signals the approximate quantity of raw silk that can be reeled. The whole weight of a single cocoon is influenced by silkworm species, rearing season and harvest conditions. Pure breeds range from 2.2 to 1.5 g, while hybrid breeds weight from 2.5 to 1.8 g. In nature, the weight of a fresh cocoon does not remain constant but instead continues diminishing until the pupae transforms into a mother and emerges from the cocoon. This weight occurs gradually as moisture evaporates from the body of the pupae and as fat is consumed during the metamorphosis process

Thickness/Weight of Cocoon Shell

The thickness of the cocoon shell is not constant and changes according to its three sections. The central constricted part of the cocoon is the thickest segment, while the dimensions of the expanded portions of the head are 80 to 90 percent of the central constricted

Lustre

It is a subjective measure of the reflection of incident light from a fibre, filament or textile material. A uniform cross- sectional area of the silk fibre coupled with filament form imparts high lustre to the mulberry, muga and tasar silk yarns due to even reflection of incident light. Eri silk is also lustrous due to its higher translucency with a uniform cross-sectional area. As per birefringence data, the molecular arrangement in mulberry silk is highly oriented whereas three non-mulberry silk fibres have lower orientation with no significant.Birefringence is the difference between the refractive indices of fibre in two directions *i.e.*, along fibre axis and perpendicular to the fibre axis

Fibre Length

Silk filaments are extremely long continuous filaments. Longer fibres are easier to process and more even yarns with high strength can be produced from them.

Silk	*Total Filament Length (m)*
Mulberry (Uni)	800-1500
Mulberry (Bi)	1000-1500
Mulberry (Multi)	300-700
Multi-bi hybrids	700-900
New hybrids	1000-1200
Tasar	650-1300
Oak Tasar	650-750
Muga	300-500

Fibre Fitness

The size of the bave, raw silk is expressed in terms of denier, which refers to the weight of a length of 450m in 0.05gm units or 9000 m weighs one gram. Filament denier is not uniform throughout the reelable length of filament. It plays an important role in determining the quality of resultant yarn and hence of the resultant fabrics. The lower the denier, the finer the silk and vice-versa. Mulberry silk is the finest, followed by eri, tasar and muga. Non-breakable filament length (NBFL) is the length of silk filament that is present continuously in the cocoon.

Fibre Fineness and Non-Breakable Filament Length of Silk Fibres

Type of Silk	*Fibre Fineness (den)*	*NBFL (m)*
Mulberry (Bi)	2-3	700-800
Mulberry (multi)	2-3	400-600
Tasar	8-12	100-250
Oak Tasar	3-5	300-450
Muga	4-7	150-250
Eri	3-4	0.05-2.0

Linear Density

Linear density is the amount of mass per unit length. The mulberry silk has relatively higher degree of order and more compact molecular packaging than non-mulberry varieties because of high glycine content and lower long chain/short chain ratio. The mulberry varieties do not exhibit pores or voids in their cross-sections whereas tasar, muga and eri silk show the presence of voids. Among the three non-mulberry silks, muga silk exhibits the highest density.

X Ray Diffraction

Silk fibroin have been studied extensively on X- ray diffraction of varieties of silk. The crystalline structure of the silk was first described in 1950s as an anti-parallel, hydrogen bonded β sheet based on the characterization of *B. mori* fibroin (Marsh, 1955) [32]. Physical parameters such as crystallinity, crystallite size, crystallite orientation is determined through X-ray diffraction. Crystalline regions in the mulberry silk yield 48 per cent, tasar silk about 39 per cent, while muga at about 40 per cent and eri about 36 per cent. It is observed from X-Ray diffractograms that mulberry silk varieties have similar distinct peaks, with better-ordered regions in comparison with the double peaks and distributed intensities as in non- mulberry silks

Crystallinity and Crystallite Size

Crystallinity is the indication of crystalline region in polymer with respect to polymer content. Silk is a semi-crystalline material. Earlier studies by X-ray diffraction analysis indicated 62 per cent -65 per cent crystallinity in cocoon silk fibroin from the silkworm, 50 per cent -63 per cent in wild-type silkworm cocoons, and lesser amounts in spider silk. According to Bhat and Nadiger (1980), non-mulberry silk varieties show large crystallite size than mulberry silks attributed to higher alanine content (Ala-Ala links) in crystalline region. Three non-mulberry silks show large crystallite size and low crystallinity than mulberry silks which indicates high elongation. The presence of amino acids with bulky side groups offers lower crystallinity in non-mulberry silks with increase in crystallinity and crystallite size from outer layers to inner layers

Moisture Regain

Silk is a very hygroscopic textile fibre. Mulberry raw silk has a moisture regain of 11 per cent, which reduces to about 9 per cent after degumming (at standard atmospheric condition, 27 °C and 65 per cent RH). Tasar silk shows highest moisture regain value followed by eri and muga for outer layers. The higher moisture regain of non-mulberry silk suggests that they may consist of higher ratio of hydrophilic and hydrophobic amino acid residues. Moisture regain of inner layers is less compared to outer layers because of compactness. Silk fibres can absorb upto 30 per cent moisture from the air without feeling damp and the fibres display 69J/g of heat of wetting from 0 °C which indicate that coupled with high regain of about 10 per cent, it offers sufficient time for the wearer to acclimatize to the change in weather. Silk fibres swell about 30 per cent of their volume under wet conditions: because of this, silk textile materials have a lower dimensional stability compared to other natural fibres. Due to this swelling action, silk fibres display partial loss of strength under wet condition.

14.5 Mechanical Properties

Tenacity and Elongation

The average tenacity of mulberry (bivoltine) and mulberry crossbreed was 3.75 g/d and 3.85 g/d respectively, while tenacity of tasar was highest *i.e.*, 4.5 g/d among the three non- mulberry silks. The tenacity of muga ranged between 3.2 and 4.95 g/d. Eri silk showed lowest tenacity value, ranging between 2.3 and 4.0 g/d. The variation in mulberry is not all that high, making eri the most non-uniform in terms of tenacity along the fibre. Wild silk possesses better elongation compared to mulberry silk. The mechanical properties of silk change substantially along the length and follow a definite trend

Elastic - Plastic Nature

Silk is considered to be more plastic than elastic because of its very crystalline polymer system. Silk fibre may be stretched from 1/7 to 1/5 of its original length before breaking and returns to its original size with losing little elasticity. Stretching disorganizes the polymer system of silk which is seen as a distortion and wrinkling or creasing of the silk textile material. There is a slow elastic recovery or creep after extension, but the silk does not regain its original length.

Breaking Extension

Elongation is defined as the length to which a fibre may be stretched before breaking. Elongation-at-break showed a higher value for all the non-mulberry silks compared to mulberry varieties. The values range between 31 per cent and 35 per cent for tasar, 34 per cent -35 per cent for muga and 29 per cent -34 per cent for eri silks, respectively. The elongation values for mulberry varieties ranged between 19 per cent and 24 per cent. The relatively high degree of extension in the case of non-mulberry silks may be attributed to the following: (1) All the non-mulberry silks contain more amino acid residues with bulky side groups than the mulberry silk varieties. This enables molecular chains in non-crystalline regions in the fibre structure to slip easily when stretched and thus show higher elongation at break. (2) Unfolding of the long fibroin chains in the amorphous regions is a result of either less orientation or less crystallinity.

Intial Module

Another important tensile parameter is the initial modulus. The initial modulus values for different varieties changes along the filament length. The initial modulus values follow a similar trend as that of tenacity, showing higher values for the inner layers. The values ranged between 53.9 and 136.8 g/d for mulberry (Bivoltine),

83.8 and 129.8 g/d for mulberry (Crossbreed), 61.3 and 100.7 g/d for muga, 61.4 and 107 g/d for tasar, and 71.1 and 107 g/d for eri silk. This definitely indicates an increase in orientation in both the crystalline and the amorphous regions, as one moves from the outer to the inner layers.

Stree - Strain Curve

Silk is a visco-elastic material and exhibits the phenomenon of stress-strain. Creep and stress behaviour of silk has been reported by Das (1996). The extension, secondary creep is higher for tasar silk compared to those for mulberry silk and stress relaxation was also found to be more in non-mulberry silks. Non mulberry silk exhibits a high initial resistance followed by substantial yielding. The reasons are (1) fewer crystalline regions and (2) imperfect crystallites or entanglements, which initially offer resistance but later give way and allow easy deformation in the amorphous regions until strain hardening occur.

Cohesion

Cohesion implies the force with which the composite filaments cohere to gather inside a strand. The measurement is done by degree of agglutination of filaments and sericin plays the role in it. Abrasion is the rubbing or scrapping off the surface of a thread. Fibre crimp increases the cohesiveness, resilience and resistance to abrasion.

Resiliency

Resiliency means that the fibre can be compressed or crushed and, on release of pressure, will tend to return to its original shape. This quality causes the fabric to be wrinkle-resistant. Silk fibres retain their shape and have moderate resistance to wrinkling. Fabrics that contain a large percentage of weighting or made from short-staple spun silk have less resilience.

Drapability

The drapability of a fabric, or its ability to hang and fall into graceful shape and folds. Silk has excellent drapability because of its elastic and resilient properties.

14.6 Dielectric and Frictional Properties

Silk fibres are insulators for electrical conduction. Therefore, under the action of friction, static electric charges tend to develop in the fibres. The moisture regain dissipates the static charges effectively; however, under low humidity conditions, static charges pose problems for silk fibre handling. Like most textile fibres, silk fibres get positive static charges. The insulation resistance and dielectric strength of silk fibres give an indication of their dielectric constant, current leakages at certain voltages, moisture content, and stability under electric fields. Electrical and dielectric properties have gained importance with applications such as moisture measurement, evenness measurement, and the use of silk fibres in the form of fibre reinforced composites as insulating materials for special applications. The electrical resistance ($R_{s,}$ Ω kg/m^2) of silk fibres is 9.8(log R_s value), which is much higher than cotton (about 7), wool (8.4), and polyester (8) fibres at 65 per cent RH (Morton and Hearle, 2008) [33]. The electrical resistance decreases with rise in humidity and temperature. The force that holds together the fibres in a spun yarn and the interlacing threads in a fabric is called as frictional force. If the friction is too low, the yarn strength will fall and dimensional stability of cloth will be reduced. Textile fibres have coefficient of friction (μ) values ranging from 0.1 to 0.8, the lower value indicating higher frictional resistance. Typical μ values for silk fibre to fibre friction is 0.26 for crossed fibres and 0.53 for parallel fibres. The higher value is because of the smooth fibrous surface coupled with crystalline regions in the fibre while this fibrillar nature makes the silk fibres a poor resistant to abrasive wear.

14.7 Thermal Properties

Silk fibre is thermally stable below 100 °C. High degree of molecular orientation of silk fibroin helps the thermal stability of the silk fibre. Yellowing begins to occur in silk fibres at 110 °C after 15 minutes of exposure. From the peaks of Differential Scanning Calorimetry (DSC) curves of silk, Nakamura *et al.* (1994) predicted that the glass transition temperature of silk is about 175 °C and silk fibre degradation begins at 280 °C with an initial weight loss starting at about 250 °C. When silk fibre is subjected to heat, no significant changes occurs in the crystalline structure but the amorphous region becomes highly oriented. Silk is a good insulator of heat among the textile fibres; the specific heat of dry silk fibre is 1.38j/gK. The thermal conductivity of mulberry silk fibre in longitudinal and transverse direction is 1.49 and 0.119 W/(mK), which indicates high orientation of fibroin molecules along the direction of fibre. Due to lower thermal conductivity and high moisture regain of silk fibres, the comfort level of wearing silken items is decreased in hot and humid areas. The heat of setting of twisted silk yarns is also carried out by utilizing the thermal behaviour of silk fibres, Steam heating of the twisted silk yarn increases the moisture level and temperature in the silk fibres.

14.8 Chemical Properties

Silk is a protein fibre; it is composed of different amino acids. The properties of proteins depend primarily on the properties of the reactive groups of their constituent amino acids combined with the properties associated with the size of the protein. Amino acids are bifunctional, *i.e.*, they contain both acidic carboxyl (coo-) and amino group (NH3+) thereby acting as Zwitterion.

Action of Water

Water does not permanently affect silk fibre. It decreases about 20 per cent in strength when wet, but regains the original strength upon drying. The amorphous regions of silk are reported to be more accessible to deterioration by water. When steeped in warm water, the fibre swells but does not dissolve. The water use for reeling should be clean to prevent the dissolved substances present in it.

Action of Acid

Concentrated hydrochloric acid dissolves the silk in one or two minutes. Non-mulberry silks are less affected by the action of acids than mulberry silk. When silk is treated with dilute hydrochloric acid, it dissolves in 1-2 minutes without losing its strength. If silk is treated with strong sulphuric acid for few minutes, then rinsed and neutralized, contraction of fibres occurs from 30-50 per cent in length and losses its lustre. This property is utilized to impart creep effects. Action of nitric acid produces a bright yellow colour on silk which can be removed by treatment with stannous chloride. Formic acid and acetic acid have no injurious effect on s silk. Exposure to weak acids results in the 'scroop effect', which is a famous finishing treatment that produces a crackling noise when such fibres are rubbed together. Shell weight loss and silk filament recovery is influenced on cooking the tasar cocoons with tartaric acid. Mechanical properties do not change with the process parameters.

Action of Alkali

Silk fibres have low resistance to alkali and are easily damaged by exposure to weak alkali at elevated temperatures. Alkali conditions hydrolyze the polypeptide bonds of fibroin from its molecular chain ends, thus degrading silk fibre rapidly. Alkali solutions cause the silk to swell, due to the partial separation of the silk polymers by the molecules of alkali. Salt linkages, hydrogen bonds, vander walls force are all hydrolysed by alkali so dissolution occurs readily. Silk fibre dissolves on treatment with strong hot alkalies such as caustic soda or potash. Ammonia and alkaline soaps dissolve only the sericin layer of silk but have no effect on fibroin. Sericin or fibroin has no effect on treatment with borax. If raw silk is steeped in lime water, the fibre swells and sericin becomes soft. Continued treatment with lime water cause brittleness in silk.

Action of Metallic Salt

Silk has great affinity for metallic salts. This characteristic is employed for the process of silk weighting, to improve draping properties. Stannic chloride is used for weighting unless the material is to be dyed black. Black silk is often dyed and weighted by using log-wood and iron salt. A moderately weighted silk contains 25-50 per cent of salt.

Effect of Organic Solvent

Effect of different organic solvents on mulberry silk: While treatment of mulberry silk with ethanol, ethyl acetate, methanol there is change in the secondary structures of silk fibroin from random coil to high strength β sheet structure and thermal stability also increases. Methanol is the best solvent to induce strength and thermal properties.

Treatment of muga silk with organic acids: Modifies the fibre chemically and increases the tensile strength. Treatment with Methanol and phenol shows higher tensile strength.

Action of Dyes

Dye-stuffs are absorbed by silk more readily and at lower temperature than any other normal fibre. Silk was dyed with vegetable dyes extracted from stems, roots, stalks, foliage, barks *etc.* of different plants in older. The selection of dye is very important as dyeing increases the value of silk. Acid dyes produce brilliant shades with good light fastness but poor to moderate wash fastness which can be improved by treatment with cationic dye fixing agent at 40-50 °C for 20 min. Dyeing of silk fabric with acid dyes improves good colour fastness and mechanical properties like count, cloth thickness and cloth weight. Reactive dyes have gained importance on account of brilliant colours, fastness properties and comparatively low cost. Treatment of muga silk fabric with acid and reactive dye results in decreased fabric weight, tensile strength, bending length but increased crease recovery of muga silk fabric.

Action of Bleaching Agent

Degumming followed by sequential oxidative and reductive bleaching on mulberry and tasar silk fabrics decreases fabric weight, thickness, tenacity, crease recovery, bending length whereas air permeability increases. Hydrogen peroxide, performic acid, peracetic acid and potassium permanganate are some of the oxidising agents use for bleaching of silk. The reducing agents used are sulphur dioxide and sodium hydrosulphite. Treatment of natural colour silk with hydrogen peroxide doesn't change the structure, the degree of crystallinity decreases a little, the breaking tendency and the breaking elongation declines, the elastic modulus didn't change significantly.

Action of Perspiration

Perspiration greatly tenders silk fibre. It becomes alkaline when deodorants containing aluminium chloride are used.

Effect of Sunlight

Sunlight disintegrates the silk. Raw silk is more resistant to light than degummed silk. The changes to mechanical strength on light ageing, with and without UV radiation has been analysed by Koremberg (2007). Yellowing has been attributed to the decomposition of tryptophan accelerated by the presence of tyrosine on light exposure.

14.9 Factors Affecting the Quality of Silk Reeling

Water

The water used in silk reeling plays an important role. It is estimated that to produce one kg. Of silk from charka, the volume of water required is about 100 liters. If the quality of water is poor, the impurities suspended in the water will adhere to the silk thereby the colour, and lustre of silk would be influenced. The water used for cooking cocoons should be colour less, odour less and transparent. The pH of water should be between 6.8 to 8.4.

Cocoon Drying

The cocoons contain live pupae which will emerge out with in 10-12 days of cocooning hence, cocoons ae stifled. Once the moth emerge out, the shell becomes of no use to reel since the adult emerges out by piercing the cocoon shell. The cocoons contain live pupae, which will emerge out with in 10-12 days of cocooning hence, cocoons are stifled. Once the month emergesout, the shell becomes of no use to reel since the adult emerges out by piercing the cocoon shell. The primary object of drying cocoons is to kill the pupae and to reduce the water content of fresh cocoons.

14.10 Methods of Stifling

1. Sun Drying

The freshly harvested cocoons are directly exposed to sunlight. In this method the harvested cocoons are spread in one layer and directly exposed to bright sun-light for several days from morning to evening until the pupae inside is killed and cocoon is found properly dried. This can be identified by shaking the cocoon which will make rattling sound.

Advantages

- ☆ This is cheapest method of stifling
- ☆ Easily available method of cocoons drying to rearers.
- ☆ These cocoons can be reeled immediately.

Disadvantages

Continuous exposure to hot sun-light, damages the silk quality due to the action of ultra violet rays.

- ☆ Some times continuous sun-light may not be possible, particularly during rainy season. Partial drying of cocoons will spoil the quality of silk.
- ☆ Wastage of silk during reeling will be more.

2. Steam Stifling Methods

- ☆ Steam stifling is the very common and easy method.
- ☆ The fresh cocoons are directly exposed to hot wet steam.
- ☆ Basket steaming
- ☆ Barel steaming
- ☆ Chamber steaming

Advantages

- ☆ Conveniently used to stifle large quantities of cocoons
- ☆ Uniform stifling of cocoons

Disadvantages

- ☆ Steam stifled cocoons cannot be reeled immediately, other wise the sericin will wet and fibroin comes is lumps while reeling
- ☆ In steam stifling pupa is killed inside and does not dry it.

3. Hot Air Conditioning

In the modified hot air conditioning chamber there are five compartments. Each chamber is different ranges of temperature.

- ☆ The first chamber is filled with cocoons are supplied of with hot air at 50°C for about 3hrs.
- ☆ The second chamber by sliding mechanism by pulling out where 70°C for about 3hrs.
- ☆ The third chamber cocoons are dropped into 75-80°C.
- ☆ The forth chamber is cocoons dried by hot air at 80-90°C temperature
- ☆ The fifth chambers are gradually exposed to hot air in the 12 hours of time and continuous process.

Advantages

- ☆ Any amount of cocoons can be easily stifled.
- ☆ Time saving, labour saving.

Disadvantages

- ☆ Cocoons of uniform size are quite suitable for this process, but Indian multi-voltine cocoons with different sizes and different qualities need different degrees of temperature for stifling.
- ☆ Costly and skilled labours are required.

14.11 Storage of Cocoons

While storing the cocoons certain precautions should be taken into consideration other wise improper storage of cocoons leads to loss due to occurance of damage to the cocoons.

Preventive measures should be taken:

1. Avoid storage of stained cocoons
2. Storage room should have good ventilation.

3. Cocoon storage place should be clean and dry. No waste cocoons should be stored in the premises.
4. Always ensure proper drainage of cocoons before storage.
5. Relative humidity in the storage room should not exceed 70 per cent.
6. Before storage of cocoons, the storage place should be thoroughly disinfected.
7. In case of long storage of cocoons, frequent turnings should be given.

Sorting

Sorting of cocoons is already done at once, once again it should be checked prior to reeling.

Deflossing

Floss is the outer most layer of cocoon which is only a protective covering and is not reelable. Prior to reeling floss has to be removed by any of the methods to avoid obstruction during reeling.

Cocoon Ridding

After deflossing the cocoons, in the process of ridding, cocoons are separated according to their sizes by simple sieves into big, small and medium size cocoons and collected in separate containers. This will facilitate the reeling operations easy.

14.12 Reeling

Reeling is a process of unwinding of silk filament from the cocoon. Reeling process is an important activity. Depending upon the required thickness (denier) of silk thread filaments from number of cocoons are combined together and reeled. An efficient reeler will maintain the fixed number of cocoons per end to produce uniform denier silk. Reeling is carried out by distinctive methods.

Process of Reeling

After boiling (cooking) and brushing, reeling started. Reeling is the extraction of continuous silk thread of desired thickness from the cocoon without any break. Before reeling, light brushing of the cocoon surface is done to find out the tip of the silk filament. Then the ends are all separated out and gathered together, this is called end picking. Generally, filaments from multiple cocoons are combined together and transferred onto the reeling or spinning machine, which then automatically unravels the cocoons and simultaneously create a single strand of silk from the cocoons.

Throughout the reeling process, the cocoons remain in boiling water. When one cocoon gets exhausted, the second one is immediately substituted to reel a continuous and uniform yarn. So the important steps in reeling are brushing, end picking, casting, binding and traversing. The individual filament must be inter-wined or twisted and cemented firmly so that the reeled filament does not break during subsequent processing of fabrics. This twisting and cementing is done with the help of a special device, called Croissure, which remains attached with the reeling machine. Croissure helps in twisting and agglutinating the have as well as it squeezes out excess water from the reeled thread to make it dry. Reeling can be done by 3 systems:

Charkha-Reeling

In this manually operated system, inferior quality or defective cocoons are reeled. Here cooking and reeling are done in one basin. Following cooking at high temperature, the soften cocoons float in basin. The reeling ends are then taken out from 4-5 cocoons and twisted around themselves with the help of charkha reel to make uniform thread. Charkha reeling machine is traditionally home-made and used mainly in villages for small scale silk production. It uses chambon type of simple croissure where threads from two reeling ends are interwinded to form few spirals. So the silk reeled by charkha is of poor quality.

Advantages

- ☆ Investment on reeling machinery is very-very less.
- ☆ Even the inferior quality cocoons can be also reeled by charka.
- ☆ No much technology is involved in this system of reeling.
- ☆ Only one reeler can reel.
- ☆ Brushing and cooking are done in the same basin.
- ☆ The cost of production is very low compared to other reeling machines.
- ☆ Involvement of labour is very less only two persons are sufficient to maintain.
- ☆ Reeler can establish a charka unit at the back yord of his own residence.
- ☆ Production per day would be around 1 kg from each charka.
- ☆ Marketing of charka reeled silk is very easy and it will be always at his door steps.

Disadvantages

- ☆ The silk produced by charka reeling is inferior quality.
- ☆ As the reeling is done in water at very high temperature which is harmful to the reeler.
- ☆ Frequently, water is to be changed and heated again, this will increase the fuel cost.

Important Points while Reeling using Charaka

- ☆ The reel should be always smooth.

- ☆ As the cooking and brushing, reeling are done in the same water. The water should be changed frequently.
- ☆ As the cooking and brushing, reeling are done in the same water.
- ☆ After cooking the cocoons thoroughly, when reeling ends are found, the temperature of water in the reeling basin should be lowered to 40 0C by adding cold water.

Cottage Basin System

In this system, cooking and reeling are done separately. Boiling water basin is used for cooking while reeling is done in a hot water basin kept near the cooking basin. Cottage basin reeling machine uses travellettecroissure. It consists of 3 pulleys to enable the thread to inter- wine around itself and also better removal of water and gum spots from the thread. Cottage basin is also attached with traverse or distributor which is a modified hot rod. After being squeezed by the croissure, the thread is further dried as it passes on the traverse before reaching the reel. So, cottage basin yields silk of good quality with uniform thickness. In India, as much as 806 tonnes silk are produced by 4000 cottage basin systems.

Advantage

- ☆ Quality of silk produced is superior to charka silk.
- ☆ Water in the reeling basin will be fresh. This will help to maintain the quality of silk.
- ☆ Production per day can be increased, as it is power operated.

Points to be Remembered by Reeler

The waste material stuck to the jette about should be removed from time to time.

- ☆ It is advisable to take a mixture of fresh and partially reeled cocoons while reeling.
- ☆ Replace the water in the reeling basin from time to time.
- ☆ Apply the lubricant to croissure pulleys for smooth running.

Filature System

In this power-driven multi- end reeling device, cooking and reeling are done separately. It is a modern device where continuous reeling is done from cocoons of superior quality. The silk produced by filature is also of superior quality because of increased cleanliness and uniformity in thickness of the fibre. However, only 8 per cent of the total silk production in India is contributed by filature.

Advantages

- ☆ Superior quality silk can be produced

- ☆ Wastage is reduced
- ☆ High productivity
- ☆ Less labour
- ☆ Arrangements are made to control the speed of the reels and tension of the threads.
- ☆ Raw silk per day will be around 800 gms of superior quality.

Automatic Reeling Machine

Bivoltine cocoons are best suited to automatic reeling machines. This machinery requires superior quality cocoons and uniform size of cocoons with less floss. Generally the multi-voltine cocoons are not fit for automatic reeling. This is particularly due to inferior quality. The silk produced from automatic reeling machine will be superior

14.13 Raw Silk Testing

Silk weaving has reached a very high standard of industrial efficiency. In fact, today a number of varieties of silk fabrics are produced on handlooms and sophisticated power looms. This requires different qualities of raw silk. In order to assist the weaving industry in the selection of the required raw silk, it must be first tested and classified. Further, the raw silk reeling industry requires well-defined standards, which can only be achieved by silk testing. As the demand for silk is global and a number of countries compete in the trade of raw silk, it is necessary that there should be industry standards for raw silk quality so as to enable buyers to purchase raw silk at internationally accepted grades. This is the reason why all raw silk produced should be classified following testing. The testing of raw silk is based on the procedure laid down by the International Silk Association (I.S.A.).

1. Quality Test

Visual Inspection

This test is conducted on the whole lot of raw silk, which has to be classified. This is carried out in an inspection room, which is well illuminated. There are three main factors that have to be tested. These are:

Uniformity

In this test, the entire lot is inspected to assess the uniformity of colour, lustre and feel. The result is not recorded in the I.S.A. method, but is classified as good, fair and inferior under the Japanese method.

General Finish

Here, considering the presence and degree of a number of defects assesses the general finish of the lot. These defects are:

i. **Re-reeling.** Gummed skeins; gummed spots on skeins; double ends; irregular traverse and partial flack of traverse.

ii. **Finish.** Tangled filament, defective lacing, filament out of place in skein (pulled filaments loose).

iii. **Arrangement.** Lacing of booking card through skeins; non-uniform skeins; wrong twisting; raised filament; streaky filament; cut ends; discoloured skeins; foreign matter on skeins; irregular skeins on book; knots on skeins; skeins or books of different types.

iv. **Damage.** Books of irregular shape; gummed books, soiled filaments; frayed skeins; insect attached skeins; musty skeins, *etc.*

The results are expressed in terms such as good, fair, poor or inferior. Outstanding defects are mentioned in the Test Certificate.

Nature

The degrees of Colour and Lustre and the Smoothness or Hand of the lot are inspected and indicated in the following manner:

i. Colour: light, medium or deep

ii. Lustre : bright, medium or dull

iii. Hand : smooth, medium or rough

2. Sample Test

After completion of the test on the whole lot, the following tests have to be conducted. To conduct these tests, it is necessary to take out 50 skeins off the lot if the skeins weigh below 120 g. Alternatively, 25 skeins are taken if the skeins weight over 120 g. Having extracted the required number of skeins, the following test is carried out:

Winding Test

Equipment required:

1. Winding frame
2. Wheels and bobbins

Number of sample skeins drawn out for this test should be 40 skeins out of 50 skeins for a lot which weighs approximately 70 g each and 20 skeins out of 25 skeins drawn out for each skein weighing approximately 140 g.

When winding is started only the top half of the sample skeins should be wound. The winding should be carried out at a predetermined speed for a specific duration. The number of breaks that occur should be counted and noted. When breaks occur, it is necessary to note the cause of each break and this should be recorded.

Average Speed and Winding Period for winding test

Size under Test	*Preliminary Winding*	*Average Speed (meter/min.)*	*Winding Period (Minutes)*	
			70 g skeins	*140 g skeins*
12 denier or finer	10	110	60	120
13-18	10	140	60	120
19-33	10	165	60	120
34-69	5	165	30	60
70 or coarse	5	165	20	40

The winding speed and duration of winding has also to be adjusted according to the denier of the raw silk being tested. For reference purposes, the different time and speed and other requirements are shown in above table.

Size Deviation Test

Equipment required:

- ☆ Sizing reel (1.125 meters in circumference)
- ☆ Balance
- ☆ Denier scale

Only a fixed length from each skein is taken for the testing. The highest degree of deviation is noted and compared with the average size deviation of the conditioned weight. When testing raw silk of 33-denier or finer in size, 200 skeins of 450 meters each are taken out by picking out four skeins out of each of the 50 sampling skeins taken from the testing lot. In the case of 34-denier and over, 400 sizing skeins are extracted by picking eight skeins of 112.5 meters in length each from the 50 sample skeins (total length 45 000 meters).

Number of Skins in a Lot and Graduation in Scale

Sizes	*No. of Skins in a Group*	*Accuracy of Scale*		*Permissible Range of Denier*
		For 1 Skin	*For Group*	
33 denier or finer	20	0.5 denier	0.5 denier	1.5 denier
33-49 denier	40	1.0 denier	2.0 denier	4.0 denier
50-99 denier	40	2.0 denier	2.0 denier	8.0 denier
100 denier or coarser	40	5.0 denier	2.0 denier	19.0 denier

Evenness Test

The test is carried out with test samples of a fixed length using a Seriplane. These test samples represent fine passages and coarse ones, divided into 3 groups

(Evenness Variation, I, II and III) according to the degree and frequency of size variations.

Cleanness Test

This test is conducted to ascertain Super Major Defects, Major Defects and Minor Defects. Each defect carries penalty points and the difference of the total penalty points from 100 gives the test result

Cleanness Defects: These are categorized into three general groups, *viz.*, Super Major Defects, Major Defects and Minor Defects .

Super Major Defects: All major defects in length or size which are ten times larger than the minimum size of Major Defects are named Super Major Defects.

Major Defects: These are divided into five as follows:

i. **Waste.** This is a mass of tangled cocoon filaments or fibers attached to the yarn.

ii. **Large slugs.** These are somewhat thickened places in the thread 7 mm and above in length, or very badly thickened places shorter than 7 mm.

iii. **Bad casts.** These appear as abruptly thickened places in the yarn due to the cocoon filaments not properly adhering to the raw silk yearn, or caused by feeding more than one cocoon filament at a time.

iv. **Very long knots.** These are knots, which have loose ends, 10 mm and over, or those made by incorrect tying of threads.

v. **Heavy corkscrews** are places in which one or more cocoon filaments are longer than the rest, and give the appearance of a very coarse and large spiral.

Minor Defects: The minor defects are subdivided into four as follows:

i. **Small slugs**, which are considerably thickened places in the thread from 2 to less than 7 mm in length, or extremely thickened places less than 2 mm in length.

ii. **Long knots** are knots, which have loose ends below 10 and more than 3 mm in length.

iii. **Corkscrews** are places in which one or more cocoon filaments are longer than the remainder, and give the appearance of a thick spiral.

iv. **Long loops** or loose ends are loops or split ends, 10 mm and above in length, when measured along the filament.

Tenacity and Elongation Test

To test the strength of the raw silk, the breaking point (g per denier) and the degree of elongation (percentage) is carried out on the Serigraph. This test is conducted in a room, which is maintained at a standard temperature of 25°C and a humidity of 65 percent.

Cohesion Test

By means of the Duplan cohesion tester, the number of frictions required to split silk thread for the purpose of examining the state of cocoon filaments sticking together, can be counted. This test is conducted in a room kept at standard temperature and humidity. Apparatus used for this test is Duplan cohesion tester.

14.14 Importance of Reeling Industry

The word 'silk' spells luxury and class. Even today, no other fabric can match it in luster and elegance. The touch of silk on the fingertip evokes the very thread of history. Over the centuries it has reigned undisputed as the queen of textiles. Mankind has always loved silk. It is the yarn extruded by a small caterpillar in a continuous filament as long as nearly one and half kilometers.

The sericulture industry is agro-based and floushing mostly in rural areas. More than 50 per cent of silk is reeled by a villager using country charka which forms the cottage industry. Silk provides much needed work in several developing and labour rich countries. India has no exception for this statement. More than 10 million farmers raise silk in China today. India is one of the first three silk producing countries provides employment and self-employment to more than five and half million people. Among this reeling sector is a part. Since silk is a material of limited production accounting for less than 0.2 per cent of the total world production of all textilefibres. In the group of natural fibres. Which comprises cotton, wool and silk, production of silk amounts for 0.3 per cent only. It is of interest to notethat despite enormous increase in the production of man-made fabric inrecent years of global silk production has maintained steady levels.

Silk was discovered in China and the earliest authentic reference to silk is to be found in the Chronicles of Chou-King (2200 B.C.). Since then, silk figured prominently in public ceremonies as a symbol of homage to the emperors and an elegant wear in festivals, marriages and important occasions. Since its discovery the importance was recognized and still the demand is increasing for this natural fibre. It is clear from the literature that silk had a prestigious place in the culture and commerce of India even in the pre-vedic times. Indian history shows clear evidences of different silk fabric use from vedic period. The word "Subhavasthothareeya" in Laxmisookthi and"Sweethambaradh are Devinanalankara Bhushithe" in Laxmisthothram indicate silk usage during those days. Further mulberry importance was evident from Ithihas and Vedas. Besides these different words of Rugveda *i.e.* Urna meaning Silk; Manurmat meaning Cloth made of silk dates back to 140 BC. According to some historians, raw silk was exported from India to Rome during the reign of Kanishka in 58 BC probably through silk road. Artisand engaged in silk production and weaving were patronized by emperors, kings Jagirdars and other rich people through Indian history. In the pre-British period, it flourished in the states of Bengal, Mysore and Kashmir.

India has a traditional reputation for a particular kind of goods from ancient times. Indian made silk textiles are popular all over the world. This includes Banaras and Surat silks with masterly brocades; the soft-as-a sigh silks of Karnatakal the tie and dye magic of the Patola from Gujarat; the Ikats of Orissa; delicate silks of Kashmir; Sheer brilliant fabrics of Bandhej;temple silks of Kanchipuram and Tanjavur in Tamil Nadu;Dharmavaram and Pochampally silks of A.P. where master craftsman blend skillfully the art of India with the smoothness of the silk yarn to produce works of supreme creativity.

14.15 Uses of Silk

Generally, 'silk' term speaks about mulberry silk. It is soft smooth,lustrous and holds a prestigious place among textile fibres and known as 'Queen of Textiles'. Silk is sold and exported as raw silk yarn. It is woven into fabrics by powerloom or handloom and sold directly as fabrics or as readymade garments. Article made of silk *i.e.*, ties, scarves, soles, shawls,furnishings and carpets have always had a good ready market. Knitted materials from silk fibres *i.e.*, socks, stocking are very costly and possess good market.

Other variety fabric materials like dupions, plain silk, deluxe, satin,chiffon, chinnons, crepe, broacades are made from mulberry silk. Cosyfurnishing materials are made from hand-spun mulberry silk. Reeling waste, bad cocoons are used to make spun silk fabrics. The waste silks are hand spun into matka, katan, feshua (jatan or Jatam, Jhut) and noil yarns. Articles made from waste silk also have a good export market.

Among non-mulberry silks tasar silk fabrics in exotic designs are produced by handlooms. They are Gicha-noil, tasar plain, cotton-tasar blend,tasar-mulberry blend, peduncle fabric. Muga silk occupies sarees and dress material production. Eri spun silk is used for dress materials and the coarse variety for making chaddar, shawls and quilts.

Trimoulters silk yarn is used as package material in pencil industry and for making talcum powder puffs. Silk is used as raw material for pre- paring sound-free gears for making precision machinery. In France 22-24denier silk is used in tyre manufacturing to have a longer life span thanrubber tyres. Parachutes are made from 13-15 denier silk fiber. These para-chutes were used in World War-II. Cosy and soft sky jackets, comforters and sleeping bags are also made from silk. The silk gut used in surgery for internal suturing is made from silk glands. The silk glands are dissected out and put in warm water and pulled at two ends to yield a fibre of uniform thickness. This protein is auto absorbable and need not be removed after wound healing. Silk grafts have been used successfully to replace cut arteries.

Silkworm can be reared in laboratory genetic studies. This insect was proved as a good laboratory tool for any kind of experiment. Lot of research work is under progress on different lines of biotechnology, genetics using silkworms in Japan.

14.16 Raw Silk Production

The synthetic fibres had only as short period of supremary. Among all natural fibres silk does not produce any skin allergies. Its luster, smooth texture, its natural affinity towards dyes, its wetting properties, its thermal properties and other physical properties are unmatched in any of the other natural or man-made fibre. Keeping in view of the demand for raw silk for various uses the Government of India took certain steps but did not capture a good share in exporting. The products like ties, scarves, chadder, carpets,furnishing are exported from India.

In India silk trade was started in 1670. Tipu Sultan established the Mysore Silk Industry and James Anderson introduced sericulture in the Madras Presidency. The government created Sericulture Departments in Karnataka (1911) and Madras (1919). Then formation of Mysore Silk Wearing Factory (1932), Mysore Spun Silks (1936), The Silk Conditioning and Testing House at Mysore (1942) and Sericulture Research Station at Berhampore (1943) helped India Sericulture Industry to make rapid development.

Then after World War-II Indian Silk Industry were at the high demand for manufacturing of parachute and powder bags. This down trend created Silk Development Directorate in 1945 and organized an All-India Silk Conference in 1947 and worked for the establishment of Central Silk Board (CSB) at Bangalore which was established in 1949 after the Parliamentary Act. The board is under the control of Ministry of Textiles, Govt. of India. The CSB functions to promote the Sericulture Industry in every line including the industry to meet global demands. It has established International Center for Training and Research in Tropical Sericulture (ICTRIS) in 1980. Later launched the National Sericulture Project (NSP)aided by World Bank and Swiss Development Co-operation. As a result of this project 32 silk worm seed production centers, 34 Basic seed farms and 5 cold storage units were established besides 80 Technical Service Centers and seven Demonstration-cum-Training Centres. CSB has set up four re-search institutes. Since then, the CSB is concentrating on different angles of Sericulture especially in reeling and improved the production of raw silk.

14.17 Scope and Limitations

Reeling industry is an ideal cottage industry. More than 50 per cent of silk is reeled by a village using country Charka and contributing to cottage industry. The system of cottage reeling is well suited to rural conditions with its low initial investment, village made simple appliances, labour with a little skill. Further the government is supplying the materials needed at subsidized rates even at the very low initial expenditure. Further almost all the states have introduced different schemes for financial, technical and material assistance. Besides this Nationalised Banks, Co-operative Societies and other funding agencies are extending loan facilities for all activities of sericulture.

India has the unique distinction of being the only country in the World which cultures all the known commercial varieties of silk *i.e.*, Mulberry, Tropical and Temperate Tasar, Eri and Muga. Out of these India holds the World monopoly in the production of Muga-the Golden Yellow Silk. The sericulture industry is one of the most labour intensive sectors of the India economy coming both agriculture and industry. Its operations pro-vide means of livelihood to a large section of the population. Of which reeling industry contributes more employment as reeler, twister, weaver, hand spinners of silk waste, traders, by product collector. Further it is turned as highly remunerative cash crop in a short span of time with minimum investment and became successful in attracting the new entrepreneurs. Among all the activities of sericulture the reeling sector gets 43 per cent share distributed at all angles it provides whole time and part-time employment to rural people of the country, majority of them belonging to economically weaker sections, Schedule Castes, Schedule Tribes as well as a divas (tribals).

The income generating sources in sericulture are open in

- ✰ Production of Reeling Cocoons
- ✰ Production of Raw Silk
- ✰ Utilization of bye-product *viz.*, Pupae and Silk waste

The demand for silk fabric is increasing since its invention. India is very poor producer of raw silk to meet international standards. In India raw silk production is overtaken by silk waste. India produces annually about180 tonnes of spun silk yarn and 130 tonnes of noil yarn besides hand spunyarn. A wide range of products made of these waste silks such as dress materials and furnishings which are having considerable export demand. Keeping in view of increasing demand India has to work out for improvement of these sectors for further more employment generation as well as export business. Silk weaving is largely undertaken on handlooms which required no license. There is need for better mechanization of handlooms. On the other hand silk waste processing and production of spun silk yarn isa scheduled industry which falls under schedule 23(4) of Industrial Development Regulation (IDR) Act, 1951 which requires a Industrial License under Ministry of Industry, Government of India, New Delhi.

Silk reeling, using country charka is of low quality but it contributes more to the raw silk produce. The reasons for low quality silk in country charka are as follows:

- ✰ No sorting of cocoons. Defective cocoons are also reeled.
- ✰ No cocoon mixing
- ✰ Improper deflossing.
- ✰ As there is no jette bout the dirt, gum spots *etc.*, can be eliminated. However, the charka has Tharpatti but cannot function like button or jettebout.

AT A GLANCE

- ☆ Reeling is to be carried with proper skill and good machines.
- ☆ The aim of cottage reeling industry is to obtain the maximum yield of raw silk from cocoons at the minimum cost of material and labour.
- ☆ No fiber (natural or synthetic) can match silk in luster and elegance.
- ☆ Sericulture is alongbased and flourishing mostly in rural areas and contributes 50 per cent of raw produced on country charka.
- ☆ Silk had a prestigious place in the culture and commerce of India inthe Pre-Vedic times. Its usage was also coined in Itihas and Vedas (Rugveda).
- ☆ India is famous for Banaras, Surat Silks, Tie and Dye, Patola, Ikats,Kashmir, Karanataka, Bandhej, Kanchipuram, Tajavur, Dharmavaram, Pochampally Silk Fabrics.
- ☆ Silk is known as 'Queen of Textiles'. It is sold as raw silk, fabric or as readymade garments.
- ☆ Reeling waste, bad cocoons are used to make spun silk fabrics. The waste silk is made into matka, katan, feshua, noil yarn which have very good export market.
- ☆ Silk is used for fabric power, package material, sound free gears,tyre manufacturing, parachutes, sky jackets, sleeping bags,confortors, suturing material.
- ☆ Silkworm is very much useful for genetic experiments in biotechnology.
- ☆ Silk trade was started in 1670 when Tipu Sultan established the Mysore Silk Industry.
- ☆ The central Government has taken lot of steps for promoting the sericulture activities, and established Central Silk Board in 1949 at Bangalore.
- ☆ CSB has setup research stations, cold storages, seed production entrées, basic seed farms, technical service centres, demonstration-cum-training centres, certificate and diploma courses.
- ☆ The government has introduced different schemes for financial, technical and material assistance.
- ☆ India is the only country producing muga the Golden Yellow Silk in the World.
- ☆ Reeling industry creates good amount of job, wage, self-employment opportunities to suit rural, economically backward, Scheduled Caste and Scheduled Tribes.
- ☆ Income generating sources of reeling are production of reeling co-coons, production of raw silk, utilization of bye-products.

- ☆ Indian raw silk cannot reach international standards.
- ☆ Reeling industry does not require any license. But spun silk yarn production unit needs a license under schedule 23(4) of IDR Act,1951.
- ☆ Silk reeled in country charka is of low quality.
- ☆ Multiend, filaure, automatic reeling units are better for good quality raw silk production.
- ☆ There is need to improve all the infrastructure facilities of reeling industry to produce quality raw silk of international standards

This low-quality raw silk hampers the export market. The main aim and objective of silk reeling is to export the raw silk for which reeling process is to be undertaken by technical hands-on standard reeling machines. Silk reeling industry equipped with improved machines like multiend reeling machines or automatic reeling machines have wide prospects. This aspect of sericulture is to be given proper attention to improve the reelingmachines, training the technical persons to operate latest machines, production of quality cocoons to reelout on automatic and semi-automatic reeling machines. There by process of reeling as well as raw silk quality improves and also gives good foreign exchange. Country charka has to be improved by providing button. The reelers are to be given training to adopt themselves for latest machines. Reeling industry must be given priority for loans by Central Silk Board (CSB), State Government departments through their directorate, NABARD, Co-operative Banks, Nationalised and Grameena Banks. Training is to be imparted by State and Central Government Schemes, such as TRISEM, D.R.D.A., Nehru Yuva Kendras.

According to the latest statistics available on income per day is Rs.45.16 in the reeling machines compared with Rs. 10.36 in country charka. Similary income that comes by selling raw reeled on country charka is very less when compared to silk reeled out on improved or semi-automatic reeling machines.

In India semi-automatic and automatic reeling machines are con-fined to certain states and practically they are not being used because the green cocoons which are available cannot suit to feed such machines. Thus, lot of importance is to be given to evolve new races to suit the latest equipment so as to get good cash returns. Reeling industry needs lot of technical persons which is now lacking. Therefore, training programme are to be launched to benefit the reelers.

References

Abbas SQ, Ali I, Niaz M, Ayesha R, Iftikhar T, (2010) new fungal records on Morus alba from Faisalabad, *Pakistan Pak. J.Bot.,* 42(1): 583-592.

Adams JR, Bonami JR. Atlas of invertebrate viruses. CRC Press, Boca Raton, FL, USA, 1991; 684.

Aiswariaya, K.K., Manjunatha, M., Mohan, Naik, I. (2007) Biology and host range of spiralling whitey. *Karnataka Journal of Agricultural Sciences.* 20(1): 149 – 152.

Aizawa K. The nature of infections caused by nuclear polyhedrosis virus. In: Insect pathology: An Advanced Treatise (Ed. by: Edward, A. Steinhaus) Volume 1, Academic Press, New York and London, 1963; 381-412.

Al-kirshi, R.A, Alimon A.R., Zulkifli, I, Zahari, M. W. and Sazili, A.Q. 2009. The Chemical Composition and Nutritive Value of Mulberry Leaf as a Protein Source in Poultry Diets. International Seminar on Animal Husbandary. Pp. 98-102

Altman GH, Diaz F, Jakuba C, Calabro T, Horan RI, Chen J, et al. Silk based biomaterials. Biomaterials. 2003;24: 401-416.

Ananthalakshmi, K.V., BAIG, M. AND SAMSON, M.V., 1992, Ann. Report, CSR and TI, Mysore, p. 53.

Anonymous. Garment and fabric terminology: A Primer. Retrieved on 28th January, 2022. From https://blog.unitexdirect.com/garment-and-fabric- terminology-a-primer/.2009.

Anonymous. Silkworm disease. Agricultural services bulletin 73/4. Food and Agriculture Organization of the United States, Rome, 1991.

Arora S, Pastorella G, Byrne B, Marsili E, O'Kennedy R. Microbial cells and biosensing: A dual approach– exploiting antibodies and microbial cells as analytical/power systems. In: Zacharis, C.K., Tzanavaras,

Arora S. Effect of printing on physical properties of muga silk fabric with reactive and acid dyes. International Journal of Home Science. 2016;2(3): 20-23.

Arti Sharma, Palvi Sharma, Jyoti Thakur, Murali S and Kamlesh Bali (2020) Viral disease of Mulberry silkworm Bombyx mori L – A review. *J.of Pharma and Phyto,* Sp9(2): 415-423

Aruga H. Polyhedral diseases in the silkworm *Bombyx mori* (L). Revue du Ver a Soie (*Bombyx mori* L.) Journal of Silkworm, Paris, France. 1957; 8: 37-41.

Attathom T, Attathom S, Kumpratueang S, Audtho M. Early detection of Grasserie disease of silkworm, *Bombyx mori* by DNA probe. In: Proceeding of 32nd Kasetsart University Annual Conference: Plant Science, Kasetsart University, Bangkok, Thailand, 1994, 257-271.

Babu AM, Jalaja Kumar S, Vineet Kumar, Sarkar A, Datta RK, (2002) Tropic failure of *Phyllactinia corylea* contributes to mildew resistance of mulberry genotypes. Mycopathologia, 156(3): 207-213.

Babu AM, Vineet Kumar, Govindaiah, (2002) Surface ultrastructural studies on the infection process of Pseudocercospora mori causing grey leaf spot disease in mulberry. Mycol. Res., 106(8) 938-945.

Babu AM, Vineet Kumar, Datta RK, (1999) Development of special hyphal branches of Phyllactinia corylea on host and non-host surface. Mycopathologia, 145(1): 29-33.

Babu KM. Silk Fibres: Structure properties and applications, In: Handbook of Natural Fibres. 2020;385- 416.

Babu KR, Ramakrishna S, Reddy YHK, Lakshmi G, Naidu NV, Basha SS, *et al.,* Metabolic alterations and molecular mechanism in silkworm larvae during viral infection: A review. African Journal of Biotechnology. 2009; 8(6): 899-907.

Bandyopadhyay U K, Santha Kumar MV, Saratchandra B (2005). Role of insecticides and botanicals in regulating whitefly (Dialeuropora decempuncta) incidence and their influence on some economic traits of silkworm (Bombyx mori L.). Ann. Plant Prot. Sci., 13 (1): 48-53.

Bandyopadhyay UK, Sahu PK, Raina SK, Santhakumar MV, Chakraborty N Saratchandra B. (2000) Studies on the seasonal incidence of the whitefly, (Dialeuropora decempuncta (Q. & B.) causing leaf curl on mulberry in relation to abiotic factors. Int. J. Indust. Entomol., 1 (1): 65-71.

Bandyopadhyay UK, Santha Kumar MV (2001-02). Studies on incidence of whitefly and its natural enemies. Annual Report, Central Sericultural Research & Training Institute, Berhampore, West Bengal, India, Pp. 44-45.

Bandyopadhyay UK, Santha Kumar MV, Das KK, Saratchandra B. (2001) Yield loss in mulberry due to sucking pest whitefly, Dialeuropora decempuncta (Q. & B.) (Homoptera: Aleyrodidae). Int. J. Indust. Entomol., 2 (1): 75-78.

Bandyopadhyay, U. K., Sahu, P. K., Raina, S. K., Santhakumar, M. V., Chakraborty, N. and Saratchandra, B. (2000) Studies on the seasonal incidence of the whitey, *(Dialeuropora decempuncta* (Q. & B.) causing leaf curl on mulberry in relation to abiotic factors. *International Journal of Industrial Entomology,* 1 (1) : 65-71.

Banerjee AN, Kotnala OP, Maulik SR. Dyeing of eri silk with natural dyes in presence of natural mordants. Indian Journal of Traditional Knowledge. 2018;17(2): 396-399.

Banerjee R, Maji, MD, Ghosh P, Sarkar A, (2009) Genetic analysis of disease resistance against Xanthomonas campestris pv. mori in mulberry (Morus spp.) and identification of germplasm with high resistance. Archives of Phytopathology and Plant Protection 42(3): 291-297.

Baqual MF, Sharma DD, Gupta VP, Chandra shekar DS, (2000) Efficacy of biocontrol agents and plant products against Pseudomonas syringae pv. mori and Xanthomonas campestris pv. mori causing bacterial blight in mulberry. Sericologia 40(3): 479-485.

Benavides, J.E. 1994. Arboles y arbustos forrajeros en América Central. Costa Rica, CATIE.

Bhat NV, Nadiger GS, Journal of Applied Polymer Science. 1980;25: 921-932.

Bhat, I., Buhroo, Z.I and Bhat, M.A., 2017, Microsporidiosis in silkworms with particular reference to mulberry silkworm (Bombyx mori L.). 2017, I, J of Ento Research, 2; 01-09.

Bhatnagar, S and P. Khurana. 2003. Agrobacterium tumefaciens-mediated transformation of Indian mulberry, Morus indica cv. K2: a time phased screening strategy. Plant Cell Rep. 21: 669-675.

Bhattacharya SS., Chakraborty N, Shantha Kumar MV, Das NK, (1994) Population dynamics of the fungivorous insect Illeis indica Timb. (Coccinellidae: Coleoptera) and its effect of predation in controlling powdery mildew Phyllactinia corylea (Pers.) Karst of mulberry (Morus alba). Biol. Control of Insects and Pests, 130-134

Bhattarjee SS, Chakraborty N, Aswani Kumar C, Saha Kundu AK, (1994) Control of the white powdery mildew, Phyllactinia corylea (Pers.) Karst. with the lady bird beetle, Illeis indica Timb. (Coccinellidae: Coleoptera). Sericologia, 34(3): 485-493.

Bilang, R.,Kloti, A., Schrott, M. and Potrykus, I. 1994. PEG-mediated direct gene transfer and electroporation. Plant Mloecular Biology Manual. *Springer, Dordrecht*pp 1-16.

Bilimoria SL. Taxonomy and identification of Baculoviruses. In: The Biology of Baculoviruses. Granados S, Ferderici BA. (Eds.), CRE Press, Florida, 1986, 37.

Biswas S, (1992) Fungicides for mulberry disease control - an appraisal. Indian Silk, August 1992, pp 26-28.

Biswas S, Das NK, Qadri SMH, Saratchandra B, (1995) Evaluating different plant extracts against three major diseases of mulberry. Indian Phytopath., 48(3): 342-346.

Biswas S, Mandal SK and Chinya PK, (1993) Development of powdery mildew disease in mulberry and its control. Sericologia 33(4): 653-662.

Biswas S, Mandal SK, Roy BN, Sen SK, (1994) Influence of host varieties and fungicides on conidial germination of Phyllactinia corylea, Indian J. Seric., 33(1): 15-18.

Biswas S, Pavan Kumar T, (1995) Pseudocercospora mori: mulberry leaf spot fungus. Indian Silk, December 1995.

Biswas S, Sengupta K, (1994) Genetic control of disease resistance in mulberry. Sericologia, 34(1): 33-42.

Biswas S, Teotia RS, Mandal SK, (1992) Some field observations on severity of powdery mildew (Phyllactinia corylea) in mulberry. Indian J Seric., 31(1): 67-69.

Biswas S, Mandal SK, Teotia RS, Nair BP, Sengupta K, (1995) Intensity of mulberry powdery mildew in West Bengal with some measures to control. Indian J. Sericulture, 34 (2): 114-117.

Boruah G. Study on mixture of selected natural dyes on mulberry silk yarn. M. Sc. (Agri.) Thesis submitted to Assam Agricultural University, Jorhat, Assam, 2016.

Boruah RK, Vishwanath. Effect of degumming followed by sequential oxidative and reductive bleaching on mulberry and tasar silk fabrics. Indian Journal of Fibre & Textile Research. 1999;24: 290-293.

Bose PC, Majumder SK, Datta RK. Effect of some growth regulators on the yield and yield attributes of mulberry leaf under rainfed conditions. Sericologia. 1995;35: 297-301.

Central Silk Board, Bangalore, Karnataka (India).

Chakraborti, S. P., K. Vijayan, B. N. Roy and S. M. H. Quadri. 1998. In vitro induction in tetraploidy in mulberry (Morus alba L). Plant Cell Rep.17: 794-803.

Chakraborty JN. Dyeing with acid dye. In: Fundamentals and practices in colouration of Textiles, 2010, 166-174.

Chakraborty N, Santhakumar MV, Bhattacarya SS, (1994) Studies of the biology of Illeis indica Timb. (Coleoptera: coccinellidae). A predator of white powdery mildew, Phyllactinia corylea (Pers.) Karst infecting mulberry leaves. Biol. Control. Ins. Pests, 75-80.

Chauhan S, Srivastava A, Singh K, (2010) Study on the incidence of powdery mildew disease in agro-climatic conditions of Lucknow region of Uttar Pradesh., ARPN Journal of Agricultural and Biological Science, 5(3): 65-67.

Checker, V.G., Chhibbar, A.K. and Khurana,P. 2012. Stress-inducible expression of barley Hva1 gene in transgenic mulberry displays enhanced tolerance against drought, salinity and cold stress. *Transgenic Res.,*21: 939–957.

Chinnaswamy, K.P. & Hari Prasad, K.B. 1995. Fuel energy potentiality of mulberry. Indian Silk 34 (4): 15-18.

Chitra C, Karnath NGK, Vasantharajan VN. Diseases of the mulberry silkworm, *Bombyx mori* L. Journal of Science and Industrial Research. 1975; 34: 386-401.

Choudhury SN. Silk and Sericulture. Directorate of Sericulture. Guwahati, 1992, 1-5.

Choudhury, P.C. 1997. Mulberry cultivation, CSRTI, Mysore. Cultivation technology for chawki mulberry garden.

Choudhury, P.C., 1972. Handbook of silkworm rearing. Agriculture Technique Manual 1, CSRTI, Mysore, Tokyo, Japan, Fuji Publishing Co. Ltd.

Choudhury, P.C., 1976. Chemical composition and quality of mulberry leaves, p.154-165. Text book of tropical sericulture.

Chowdhury SN. Eri silk industry. Directorate of Sericulture and Weaving, Assam, 1982, 29-44.

Chowdhury SN. Muga silk industry. Directorate of Sericulture and Weaving, Assam, 1981;24-25.

Connelly JT, Nugen SR, Borejsza-Wysocki W, Durst RA, Montagna RA, Baeumner AJ. Human pathogenic Cryptosporidium species bioanalytical detection method with single oocyst detection capability. Analytical and Bioanalytical Chemistry. 2008; 391: 487-495.

Cook JG. Handbook of Textile Fibers, 1st Ed, Abington Hall, England. Woodhead Publishing Ltd, 1959, 159p.

Craig CL, Hsu M, Kaplan D, Pierce NE. A comparison of the composition of silk proteins produced by spiders and insects. International Journal of Biological Macromolecules. 1999;24: 109-181.

Crighton JS. Silk: A study of its degradation and conservation, In conservation science in the UK, Tennent,

Croci L, De Medici D, Morace G, Fiore A, Scalfaro C, Beneduce F, *et al.,* Detection of hepatitis A virus in shellfish by nested reverse transcription-PCR. International Journal of Food Microbiology. 1999; 48(1): 67–71.

CSRTI, 1996. Sericulture: New technologies, Mysore.

Dandin SB, Kumar R. Evaluation of Mulberry Genotypes for Different Growth and Yield Parameters. Genetic Resources of Mulberry and Utilization (1st edition) 1989.

Dandin, S. B., Jayaswal, J. and Giridhar, K. (2000) Hand Book of Sericulture Technologies.

Das S. Application of natural dye on silk. Colourage. 1992;39(9): 52-54.

Das S. Studies on Tasar Silk. Ph.D thesis submitted to IIT, Delhi, 1996.

Dasgupta, M.R., 1950, Diseases of Silkworm. In Monograph on Cottage Industries.

Datta, R.K. 2000. Mulberry cultivation and utilisation in India. Proceedings of the electronic conference. de Almeida, J.E. & Fonseca, T.C. 2000. Mulberry germplasm and cultivation in Brazil. Proceedings of the electronic conference.

David, B.V., Ramamurthy, V.V. (2011) Elements of Economic Entomology 6th Edition, Namrutha Publications, Chennai, Tamil Nadu.

Dayakar Yadav BR, Sukumar J, (1987) Occurrence of a new stem- blight and collor-rot disease of mulberry from India. Sericologia, 27(2): 205-206.

Devaiah, M.e, 1973, Pathological and Ecological studies on the pebrine disease of the silkworm Bomyxmori L. Ph.D. Thesis" UAS, Bangalore, p. 175.

Dhiraj K, Kumar RV. Application of foliar nutrients to increase productivity in sericulture. Journal of Entomology; 2011. ISSN: 1812-5670 DOI: 10.3923/je

Dhiraj K, Kumar RV. Application of foliar nutrients to increase productivity in sericulture. Journal of Entomology. 2011, 1-12

Du H, Li G, Dai H, Zhou S, (1987) Studies on the biological characteristics of mulberry blight strains (Pseudomonas syringae pv. mori). Sericologia, 27(2): 183-188.

Dutta, Ghosh MK, Borah A, Bindroo BB, (2012) Development of Disease Forecasting Model for Leaf rust of Mulberry (Morus alba L.) of Dimapur of North East India. Nature and Science 2012; 10(8) 66-69.

Engelmann, F. 2000. Importance of cryopreservation for the conservation of plant genetic resources. In F.Engelman and H. Takagi (Eds) pp. 8-20. Cryopreservation of tropical plant germplasm. Current research progress and application. IPGRI, Rome, Italy.

FAO. 1988. Mulberry cultivation. FAO Agricultural Services Bulletin No. 73/1, Rome. 127pp.

FAO. 1990. Sericulture training manual. FAO Agricultural Services Bulletin, No. 80, Rome. 117pp.

FAO. 1993. Possibility of combined utilisation of Morus alba and Trifolium subterraneum in the Tuscan Maremma (Italy), p. 206-209. By P. Talamucci & A. Pardini. REUR Technical Series No. 28. Rome, FAO.

FAO. 1999. Mulberry, an exceptional forage available almost worldwide, by M.D. Sánchez. World Animal Review, 93(2): 36-46.

Fujii-Kawata I, Miura KI. Segments of genome of viruses containing double-stranded ribonucleic acid. Journal of Molecular Biology. 1970; 51(2): 247-253.

Fujiwara, T. 1984. Studies on the mass pebrine inspection of mother moth. Technical Bulletin of Sericulture Experiment Station (Japanese), 120: 113- 160.

Ganga,G and Sulochana Chetty,J.2005. An introduction to sericulture. Oxford & IBH Publishing Co.Pvt. 2.58.

Gangwar SK, Pratheesh Kumar PM, Elangovan C, Qadri SMH, (1998) Response of mulberry varieties to leaf rust (Cerotelium fici) as influenced by fungicides. Sericologia, 38(4): 661-666.

Gangwar SK, Qadri SMH, Maji MD, Pratheesh Kumar PM, Saratchandra B, (2000) Evaluation of fresh plant extracts for the control of mulberry powdery mildew. Indian J. Seric., 39(1): 76-78.

Gani MU, Chouhan S, Babulal RG, Khan G, Kumar NB, Saini P, *et al., Bombyx mori* nucleopolyhedrosis (*Bm*NPV): Its impact on silkworm rearing and management strategies. Journal of Biological Control. 2017; 31(4): 189-193.

Geetha, T., Mahalingam, C.A. and Murugan, N. (2015) Biology of mulberry thrips, (*Pseudodendrothrips mori* Niwa) under Tamil Nadu climatic condition *International Journal of Agricultural Science and Research,* 5(2): 19-26.

Geethabai, M., PAT1L, es. and Kasturi BA1, A.R., 1985, Seriologia, 25 : 297.

Ghosh L, Alam MS, Ali MR, Shohael AM, Alam F, Islam R, (2003) Changes in some biochemical parameters of mulberry (Morus sp.) leaves after infected with leaf spot disease. Online Journal of Biological Sciences 3(5): 508-514.

Ghosh L, Neela FA, Mahal MF, Khatun MJ, Ali MR, (2012) Effect of various factors on the development of leaf spot disease in mulberry J. Environ. Sci. & Natural Resources, 5(1): 205 - 209, 2012

Gohl EPG, Vilensky LD. Textile Science, Edu 1, 2nd edition, CSB Publishers and Distributors, New Delhi, 1987, 31.

Gongfu, Zhong; Zengqi, Wang & Houshui, Wu. 1997. Land-water interactions of the dike-pond system. Namur, Belgium, Presses universitaires de Namur. 130 pp.

Gopinathan KP. Biotechnology in sericulture. Current Science. 1992;63(3): 282-287.

Gould AB and JH. Lashomb. (2005) APS Feature Story: Bacterial Leaf Scorch of Shade Trees. Online: http://www.apsnet.org/online/feature/bls/

Govindaiah, Gunasekhar V, Gowda P, Thiagarajan V, (1994) Field evaluation of fungicides against Phyllactinia corylea causing powdery mildew in mulberry (Mors alba L.). Indian J. Seric., 33(2): 160-162.

Govindaiah, Gupta, V.P., Sharma, D.D., Rajadurai, S. and Naik, V.N. (2005) Mulberry crop protection. Central Silk Board Publication, Bangalore.

Govindaiah, Sengupta K, Sharma DD, Gargi, Gunasekhar V, (1989) A new leaf spot disease of mulberry caused by Myrothecium roridum Tode ex fr. Curr. Sci., 58(7): 398.

Govindaiah, Sharma DD, Chaudhury PC, Mallikarjuna B, Madhava Rao YR, (1990) Incidence of fungal diseases of mulberry under different agronomical practices, Sericologia, 30(2): 237-242.

Govindaiah, Sharma DD, Sengupta K, Gunasekher V, Suryanarayana N, Madhava Rao, YR, (1989) Screening of mulberry varieties against major fungal diseases. Ind. J. Seric., 28(2): 207-213.

Govindaiah, Sharma DD, Singhal BK, Sengupta K (1990) Fusarium pallidoroseum Cooke Sacc. - a new pathogen causing leaf blight in mulberry Morus alba L. Indian Journal of Sericulture 29(2): 291-292

Govindaiah, Philip T, Sengupta K, Giridhar K, Suryanarayana N, (1990) Incidence of leaf spot disease in relation to epidermal and cuticle thickness in some indigenous varieties of mulberry. Sericologia, 30(2): 257-260

Graham RI, Rao S, Possee RD, Sait SM, Mertens PP. Detection and characterization of three novel species of reovirus (Reoviridae), isolated from geographically separate populations of the winter moth *Operophtera brumata* (Lepidoptera: Geometridae) on Orkney. Journal of Invertebrate Pathology. 2006; 91: 79-87.

Granados RR, Lawler KA. *In vivo* pathway of *Ac*NPV invasion and infection. Journal of Virology. 1981; 108(2): 297-308.

Grice KRE, Beilharz DV, Shivas RG, (2006) First record of Pseudocercospora mori causing grey leaf spot on mulberry in Australia. Australasian Plant Disease Notes, 1, 9–10.

Gulrajani ML. Bleaching of silk, Silk Dyeing, Printing and Finishing. Department of Textile Technology, IIT, New Delhi, 1989, 1-364.

Gulzar P, Tanki TN, sahaf KA, Munshi NA, Shahzad AZMA, Raja TA, (2009) In vitro evaluation of various botanical extracts against Fusarium pallidoroseum (Cooke) Sacc. - the causal pathogen of twig blight of mulberry. Ind. J. Seric. 48 (2): 133-137.

Gunasekhar V, Govindaiah, Datta RK, (1992) A new leaf blight of mulberry (Morus spp.) caused by Alternaria alternata in India. Indian J. Seric., 31(2): 131-134.

Gunasekhar V, Govindaiah, Himantharaj MT, (1995) Efficacy of fungicides in controlling mulberry leaf rust caused by Cerotelium fici. Indian J. Seric., 34 (1): 60-62.

Gunasekher V, Philip T, Govindaiah, Sharma DD, Nagaraj B, Datta RK, (1994) Seasonal occurrence of foliar fungal and bacterial diseases of mulberry in South India. Indian Phytopath., 47(1): 72-76.

Gupta VB, Rajkhowa R, Kothari VK. Physical properties of Indian silk fibres. Indian Journal of Fibre & Textile Research. 2000;25: 14-19.

Gupta VP, (1999) On the occurrence of green mosaic disease on mulberry in India. Indian Phytopath., 52(2): 154-155.

Gupta VP, (2000) Biological Control of mulberry diseases- The current status. In: Diseases of plantation crops, spices, beetle vine and mulberry. MK Dasgupta (Ed.), Visva Bharati.,Palli- Siksha Bhavana, Dept. of Plant Protection Sriniketan, West Bengal, India, 145-148.

Gupta VP, Raju HV, Vineet Kumar, Govindaiah (1998). Surface ultrastructure of infection process of Alternaria alternata and Fusarium pallidoroseum on mulberry.Archives of Phytopathology and Plant Protection 31(5): 429-434

Gupta VP, Sharma DD, Govindaiah, Chandrashekar DS, (1999) Soil solarization for the control of nursrery diseases in mulberry. Indian J. Seric. 38(1): 44-47.

Gupta VP, Sharma DD, Rekha M, Chandrashekar DS, (1999) Integration of Trichoderma pseudokoningii with agrochemicals for disease management and plant development in mulberry. Arch. Phytopath. Pflanz. 32: 521-529.

Gupta VP, Tewari SK, Datta RK, (1995) Surface ultra-structure studies on ingress and establishment of Pseudomonas syringae pv. mori on mulberry leaves J. Phytopathol., 143: 415-418.

Gupta VP, Tewari SK, Govindaiah, Bajpai AK, Datta RK, (1995) Observations on the surface ultra-structure of conidal stage of Cercospora moricola and its infection process in mulberry. Sericologia, 35(1): 123-128.

Harrison R, Hoover K. Baculovirus and other occluded insect virus. Insect pathology. 2012; 1: 73-132.

Hattopadhayay S, Maji MD, Pratheesh Kumar PM, Das KK, Saratchandra B, (2002) Response of mulberry brown leaf spot fungus Myrothecium roridum to different plant extracts. Int. J. Indust. Entomol., 5(2): 183-188

Hong HK, Woo SD, Choi JY, Lee HK, Kim MH, Je YH, *et al.,* Characterization of four isolates of *Bombyx mori* nucleopolyhedrovirus. Archives of Virology. 2000; 145(11): 2351-61.

Hong SK, Kim WG, Sung GB, Nam SH, (2007) Identification and distribution of two fungal species causing sclerotial disease on mulberry fruits in orea. Mycobiology, 35(2): 87-90

Hopkins DL (1989). Xylella fastidiosa: Xylem limited bacterial pathogen of plants. Annual Review of Phytopathology 27: 271-290.

Horton HM, Burand JP. Saturable attachment sites for polyhedron-derived baculovirus on insect cells and evidence for entry via direct membrane fusion. Journal of Virology. 1993; 67: 1860-1868.

Huo, Yongkang. 2000. Mulberry cultivation in China. Proceedings of the electronic conference.

Iijuka E, Itoh H. International Journal of Wild Silkmoth Silk. 1997;3: 37.

Iizuka E, Itoh H. Physical properties of eri silk. International Journal of Wild Silkmoth and Silk. 1997;3: 37-42.

Iizuka E. Silk thread: mechanisms of spinning and its mechanical properties. Journal of Applied Polymer Science, Applied Polymer Symposium. 1985;41: 173-185.

Illahi I, Nataraju B. Prevalence of nuclear polyhedrosis in mulberry silkworm, *Bombyx mori* L. in Jammu and Kashmir. Indian Journal of Sericulture. 2007; 46(1): 43– 48.

Inoue H, Ayuzawa C, Kawamura A. Effect of high temperature on the multiplication of Infectious Flacherie Virus in the Silkworm *Bombyx mori*. Applied entomology and zoology. 1972; 7: 155-160.

Irfan I, Aftab S, Mittal V Dhar A (2012). Sooty mold of mulberry – First report from Kashmir. Journal on New Biological Reports 1(2): 38-41.

Irfan I, Mittal V, Ramegowda GK, Dhar A, Khan MA, (2011) Occurrence of major foliar diseases of mulberry under temperate climatic conditions of Kashmir, India. International Journal of Science and Nature, 2(1): 51-54.

Ishihara R, Fujiwara T (1965). The spread of microsporidiosis in a colony of silkwom, Bombyx mori, L. J. Invertebr. Pathol., 7: 126-13.

Ishikawa S. Respiratory enzymes in the midgut of the cytoplasmic polyhedrosis infected silkworm, *Bombyx mori*. Journal of Sericulture Science, Japan. 1958; 27: 99- 103.

Jalaja S. Kumar, Sarkar, A. & Datta, R.K. A breakthrough in mulberry breeding in sustainable cocoon production. In Global silk scenario - 2001, Proceedings of the International Conference on Sericulture, p.242-247. Oxford and IBH publishing Co. Pvt. Ltd.

Janaki Ammal, E.K. 1960. The effect of Himalayan uplift on the genetic composition of the flora in Asia. JIBS, 39(3): 327-333. Cited by Sastry, 1984.

Jefferson, R.A., Kavanagh Ta and Bevan, M.W.1987. GUS fusions: â-glucuronidase as a sensitive and versatile gene fusion marker in higher plants. *EMBO J.,* 6: 3901–3907.

Jolly, M.S. 1987. Appropriate sericulture techniques, Mysore, ICTRETS.

Joshi RP, Raja IA. Polymerase chain reaction based detection of Grasserie virus, *Bm*NPV in Silkworm, *Bombyx mori*. Biosci. Biotech. Res. Comm. 2016; 9(3): 471-474.

Kanke E, Iwasita Y, Iwasaki K. Change in the number of airborne bacteria in a rearing room during the work of feeding mulberry to silkworm *Bombyx mori* L. Bulletin Coll. Agric. Utsunomiya Uni., Japan, 1987; 13(2): 21-26.

Kannan S. Role of foliar fertilization on plant nutrition in Crops as enhancers of Nutrient use, 1990.

Katsumata, 1972. Mulberry species in West Java and their peculiarities. J. Sericultural Sci. Japan, 42(3): 213-223. Cited by Sastry, 1984.

Katsumata, 1973. Mulberry species in South Vietnam. J. Sericultural Sci Japan, 42(1): 81-88. Cited by Sastry, 1984.

Kawakami K. Illustrated working process of new bivoltine silkworm rearing technology. JICA, D.S.T. Project Report, Central Sericultural Research and Training Institute, Mysore, 2001, 11-19.

Kawakita H, (1978) On the mycoplasma like organisms in the cultural tissues of dwarf diseased mulberry and kanamugura. J. Seric. Sci. Japan, 47(6): 465-471.

Kawase S, Hashimoto Y, Nakagaki M. Characterization of infectious flacherie virus of the silkworm *Bombyx mori* L. Journal of Sericultural Science of Japan. 1980; 49: 477-484.

Keddie BA, Aponte GW, Volkman LE. The pathway of infection of *Autographa californica* nuclear polyhedrosis virus in an insect host. Science. 1989; 243(4899): 1728- 1730.

Khan MR, Tsukada M, Gotoh Y, Morikawa H, Freddi G, Shiozaki H. Physical properties and dyeability of silk fibers degummed with citric acid, Bioresource Technology. 2010;21(101): 8439-8445.

Khurad AM, Kanginakudru S, Qureshi SO, Rathod MK, Rai MM, Nagaraju J. A new *Bombyx mori* larval ovarian cell line highly susceptible to nucleopolyhedrovirus. Journal of Invertebrate Pathology. 2006; 92(2): 59-65.

Kim KH, Inoue S, Kawanami M, Furuya N, Matsuyama N, (l999) Antagonistic action of Bacillus sp. AB89 to Phloeospora maculans, causal agent of mulberry leaf spot, J. Fae. Agr., Kyus hulJniv.,43 (3·4): 337- 342

Kim, H. R., K. R. Patel and T.A. Thorpe. 1985. Regeneration of mulberry plantlets through tissue culture. Botany Gazette. 146: 335-340.

King AMQ, Adams MJ, Carstens EB, Lefkowitz EJ. Virus taxonomy: ninth report of the international committee on taxonomy of viruses. Elsevier, Amsterdam, the Netherlands, 2012, 1327.

Kitahara, N. 1999. Utilisation of fodder trees for the production of milk and meat (3). Livestock Res., 53(9): 969-972. (in Japanese)

Kobayashi M, Inagaki S, Kawase S. Effect of high temperature on the development of nuclear polyhedrosis virus in silkworm *Bombyx mori.* Journal of Invertebrate Pathology. 1981; 38(3): 386-394.

Koczula KM, Gallotta A. Lateral flow assays. Essays in Biochemistry. 2016; 60: 111-120.

Kolluru VC, Girish KR, Attipalli RR. Biochemical response to drought stress in mulberry (Morus alba L.): Evaluation of proline, glycine betaine and abscisic acid accumulation in five cultivars. Acta Physiol. Plant. 2009; 31: 437-443.

Koremberg C. The effect of ultraviolet-filtered light on the mechanical strength of fabrics. The British Museum Technical Research Bulletin. 2007;1: 23-27.

Kostka SJ, Tatta TA, Sherald JL, Hurtt SS (1986) Mulberry leaf scorch, new disease caused by a fastidious, xylem-inhabiting bacterium. Plant Disease 70: 690-693.

Krishna Prasad KS, Siddaramaiah AL, (1979) Studies on the development of powdery mildew disease of mulberry in Karnataka. Ind. J. Seric., 18: 9-13.

Krishnamohan, Reddy B. Metabolic modulation of fatigue with special reference to lactate and ammonia metabolism in different skeletal muscles of frog. Ph.D., thesis submitted to Sri Venkateswara University, Tirupati, India, 1986.

Krishnaswami S, Narasimhanna MN, Suryanarayan SK, Kumararaj S. Sericulture manual-2 silkworm rearing, FAO, Rome, 1979, 1-131.

Kubomura Y, Nakayama K, (1975) On the overwintering of the causal pathogen of a bacterial blight of mulberry Pseudomonas mori (Boyer et Lambert) Stevens in dead leaves of mulberry. J. Sericult. Sci. Japan, 44(4): 287-293.

Kumari KV, Balavenkatasubbiah M, Rajan RK, Himantharaj HT, Nataraj B, Rekha M. Influence of temperature and relative humidity on the rearing performance and disease incidence in CSR hybrid silkworms, *Bombyx mori* L, International Journal of Industrial Entomology. 2001; 3(2): 113-116.

Kunz RI, Brancalhao RMC, Ribeiro LFC, Natali MRM. Silkworm sericin: properties and biomedical applications. Biomed Research International, 2016, 1-19.

Lal, S., Gulyani, V. and Khurana, P. 2008. Over expression of *HVA1* gene from barley generate tolerance to salinity and water stress in transgenic mulberry (*Morusindica*). *Transgenic Res.,* 17: 651–663.

Lee JW. Effects of foliar sprays of kinetin on the growth of mulberry tree and the metabolic activities in the leaves. Korean J. Seric. Sci. 1984; 26: 1-20.

Lee YW. Characteristics of the cocoon. In: Silk reeling and testing manual. FAO, Agricultural Services Bulletin. 1999;136: 3-14.

Lekha G, Gupta T, Awasthi AK, Murthy GN, Trivedy K, Ponnuvel KM. Genome wide microarray based expression profiles associated with *Bm*NPV resistance and susceptibility in Indian silkworm races of *Bombyx mori*. Genomics. 2015; 106(6): 393-403.

Lekha G, Gupta T, Awasthi AK, Murthy GN, Trivedy K, Ponnuvel KM. Genome wide microarray-based expression profiles associated with BmNPV resistance and susceptibility in Indian silkworm races of Bombyx mori. Genomics. 2015; 106(6): 393-403.

Li M, Chen X, Wu X, Man N, Jin W, Lu X. Genome Analysis of the *Bombyx mori* Infectious Flacherie Virus Isolated in China. Agricultural Sciences in China. 2010; 9(2): 299-305.

Li X, Wang XR. Effects of hydrogen peroxide treatment on structure and properties of natural color silk. Advanced Materials Research. 2013;821-822: 90-93.

Lo YM. Clinical Applications of PCR. Methods in Molecular Medicine, volume 6, Humana Press, 1998.

M Younus Wani, MR Mir, MF Baqual, Khanday Mehraj, TA Bhat and S Rani. 2017. Role of foliar sprays in sericulture industry. *J. of Pharm. and Phyto.* 6(4): 1803-1806. Maharashtra, India. *Biolife,* 4(1): 184-188.

M. Colloidal textile dye-based dipstick immunoassay for the detection of nuclear polyhedrosis virus (*Bm*NPV) of silkworm, *Bombyx mori* L. Journal of Invertebrate Pathology. 1994; 63: 135-139.

Maeda S, Watanabe H. Immunofluorescence observation of the infection of densonucleosis virus in the silkworm, *Bombyx mori*, Japanese Journal of Applied Entomology and Zoology (Japan). 1978; 22: 98-101.

Magnoler A. Effects of a cytoplasmic polyhedrosis on larval and post larval stages of the gypsy moth, *Porthetria dispar.* Journal of Invertebrate Pathology. 1974; 23: 263-274.

Mahale G, Naikwadi S. Effect of acid dyes on colour fastness properties of silk fabric. The Pharma Innovation. 2019;8(10): 169-172.

Maji MD, Chattopadhyay S, Pratheesh Kumar PM, Raje Urs S, (2003) Evaluation of mulberry phylloplane bacteria for biological control of powdery mildew of mulberry caused by Phyllactinia corylea. J Mycopathol. Res., 41 (2): 197-200.

Maji MD, Chattopadhyay S, Pratheesh Kumar PM, Saratchandra B, (2004) Evaluation of mulberry phylloplane bacteria for biocontrol of Myrothecium leaf spot of mulberry caused by Myrothecium roridum. *Agric. Sci. Digest.* 24(4): 252-255.

Maji MD, Chattopadhyay S, Pratheesh Kumar PM, Saratchandra B, (2007) Evaluation of some ethanobotaniclal plant extracts for fungitoxicity against Myrothecium roridum. *International J. Industr. Entomol.*, 14(2): 75-80.

Maji MD, Qadri SMH, (1999) Bacterial diseases of mulberry (Morus spp.). Sericologia, 39(1): 1-7

Maji MD, Qadri SMH, Pal SC, (1998) Xanthomonas campestris pv. mori- a new bacterial pathogen of mulberry. Sericologia, 38(3): 519-522.

Maji MD, Qadri SMH, Pal SC, (2003) Evaluation of mulberry phylloplane microorganisms for biological control of bacterial leaf spot of mulberry caused by Xanthomonas campestris pv.mori Indian J. Agric. Res., 37 (4): 307 – 309.

Majumder SK, Bose PC, Banerjee ND, Pratheesh Kumar PM, (2003) Ascorbic acid, an index of fungal infection of mulberry leaf. Sericologia, 43(1): 119-122.

Mallet F, Oriol G, Mary C, Verrier B, Mandrand B. Continuous RT-PCR using AMV-RT and Taq DNA polymerase: characterization and comparison to uncoupled procedures. Biotechniques. 1995; 18(4): 678- 687.

Mamatha M, Balavenkatasubbaiah M. Progressive morphological changes in the susceptible and tolerant breeds of silkworm infected with *Bm*IFV. Indian Journal of Sericulture. 2008; 47(1): 64 -69.

Mani, M. (2010) Origin, introduction, distribution and management of the invasive spiralling whiteûy *Aleurodicus dispersus* Russell in India. *Karnataka Journal of Agricultural Sciences,* 23(1): 59-75.

Manna SS, Sengupta D, Rao GS. Measurement and comparison of cohesion of silk filaments in silk strands. Indian Journal of Textile Research. 1989;14: 191-192.

Marsh RE, Corey RB, Pauling L. An investigation of the structure of silk fibroin. Biochimica Biophysica Acta. 1955;16: 1-34.

Martinex-Zubiaur Y, Abreu MP, Hernandez MCP, Sihler W, Falcao R, Ribeiro BM, Lobo de Souza M. First record of a *Bombyx mori* nucleopolyhedrovirus (*Bm*NPV) isolate from Cuba. Journal of Current Research. 2016; 8: 35766-35770.

Mehla, R.K., Patel, R.K. & Tripathi, V.N. 1987. A model for sericulture and milk production. Agricultural Systems, 25: 125-133.

Minimizawa K. Studies on the mechanism of localised sex differentiation in mulberry. VI. NAA treatment. J. Seric. Sci. Japan. 1974; 42: 146-149

Mitani, 1920, Cited in Pebrine and its Control (Ed. Jolly, M.S), 1986. CSR and TI, Mysore, p.34.

Mitsueda T, (1963) Some observations on the mycelia of mulberry rust, Aecidium mori BARCLAY, in host tissue. *Journal of Sericultural Science of Japan,*.32(3): 3: 134-14.

Miyajima S. Effect of high temperature on the incidence of viral flacherie of the silkworm, Bombyx mori L. Research Bulletin of Aichi Ken Agricultural Research Center. Series D. Sericulture (Japan). 1978; 9: 45-50.

Monreal CM, De Rosa, Mallubhotla SC, Bindraban PS, Dimkpa C. The Application of Nanotechnology for Micronutrients in Soil-Plant Systems. VFRC (Virtual Fertilizer Research Center) Report, Washington, D.C. 2015, 44.

Morton WE, Hearle JWS. Physical Properties of Textile Fibres, 4th Ed. Cambridge, England, Woodhead Publishing Ltd. And CRC Press LLC, 2008, 1-796.

Mukherji, N.G., 1912, Handbook of Sericullure. Bengal 'Srcretariat Book Depot, Calcutta, India.300.

Munshi NA, Zargar MA., Tanki TN, Lone AH, (1991) Screening of some mulberry varieties against Phleospora leaf spot disease of mulberry under Kashmir condition. Sericologia, 31(4): 719-723.

Murashige, T. and F. Skoog. 1962. A revised medium for rapid growth and bioassays with tobacco tissue cultures. Physiologia Plantarum.15: 473-497.

Naegeli, 1857, Cited in Pebrine Disease of Silkworm - A Technical Report (Ed. Tatsuke,K., 1971), Overseas Technical Co-operation Agency, Tokyo, Japan.

Nakamura S, Magoshi J, Magoshi Y. Thermal properties of silk proteins in silkworms. ACS symposium Series. 1994;544: 211-221.

Narahari R, Krishna BV, Chaluvachari M, Bongale UD. Studies on the effect of commercial foliar spray formulations on growth yield and quality of mulberry. National Seminar on Mulberry Sericulture Research in India. 2001, 76-77.

Narayanaswamy T. K and Sunil Kumar T (2018) 'A Review Paper On Pebrine Disease Of Mulberry Silkworm Bombyx moriL.', International Journal of Current Innovation Research, 4(5), pp. 1220-1223.

Narendra Kumar, J.B. and Shekhar, M.A. (2012) Management of giant African snail infesting mulberry. *Annadata,* 7(2): 23.

Narendra Kumar, J.B., Shekhar, M.A and Qadri, S.M.H. (2011) Giant African snail in Mulberry: Physiology and management. *Indian Silk,* 1 (12): 4-5.

Narendra Kumar, J.B., Sreenivas, B.T., Divya, S.H. and Shekhar, M.A. (2013) New white y species *Dialeuropora decempuncta* Quaintance (Homoptera: Aleyrodidae) infesting mulberry in Karnataka. *Insect Environment,* 19(2): 108-109.

Narendra Kumar, J.B., Sreenivas, B.T., Shekhar, M.A., Divya, S.H. and Mathur, V.B. (2012) Millipedes infesting mulberry in Mandya district. *Indian Silk,* 3 (51 old) (1): 4-6.

Nataraju B, Dutta RK, Baig M, Balavenkatasubbaiah M, Samson MV, Shivaprasad V. Studies on the prevalence of NPV in sericultural area of Karnataka. Indian Journal of Sericulture. 1998; 37(2): 154-158.

Niino, T. 2006. Developments in plant genetic resources cryopreservation technologies. Proc. Of the APEC workshop on effective gene bank management in APEC member economies. Suwon, Korea, pp. 197-217.

Nishitha Naik V, Sharma DD, Ma!a VR, Rekha M, Govindaiah, Gupta VP, (2008) Prediction of leaf rust disease in mulberry for timely management. Sericologia, 48(2): 195-202.

Nomani KR, Mukherjee PK, Krishnaswami S, (1970) Studies on the effect of feeding multivoltine silkworm (Bombyx mori L.) larvae with mildew affected mulberry leaves. Ind. J. Seric. 64(8). 49-52.

Noor-Ul-Din S. Seasonal variation in nutrient components of mulberry leaf and their uptake in response to split applications of NPK under temperate conditions. Ph.D. D thesis submitted to Sher-e-Kashmir University of Agricultural Sciences & Technology of Kashmir. 2012, 121-122

Nuanchai K, Wilaiwan S, Prasong S. Effect of different organic solvents and treatment times on secondary structure and thermal properties of silk fibroin films. Current Research in Chemistry. 2010;2(1): 1-9.

O'Reilly DR, Miller LK, Luckow VA. Baculovirus Expression Vectors. Laboratory Manual, W. H. Freeman and Company, New York, 1992, 47-98.

Ohnishi T. Influence of gibberellin treatment on the growth of mulberry leaves under different conditions of soil water content. J. Seric. Sci. Japan. 1986;55: 289- 292.

Okamoto S, Kimura M. Journal of the Society of Textile and Cellulose Industry Japan. 1953;9: 284.

Ottman, Y.1987. Rediscovering the realm of fruiting mulberry varieties. *Fruit Varieties Journal,*41(1): 4–7.

Oviedo, F.J.; Benavides, J.E. & Vallejo, M. 1994. Evaluación bioeconómica de un módulo agroforestal con cabras en el trópico húmedo. In: J. Benavides, Árboles y arbustos forrajeros en América Central, p. 601-6290. Vol. I. Costa Rica, CATIE, C.R. p.601-629.

P.D. (Eds.). Reviews in Pharmaceutical and Biomedical Analysis, Bentham Science Publishers, Sharjah, UAE, 2010, 63-75.

Padaki NV, Das B, Basu A. Advances in understanding the properties of silk, In: Advances in Silk Science and Technology. Woodhead Publishing, Elsevier, 2015, 3-15.

Paillot A. Traite des maladies du ver a socie. Doin, Paris, 1930, 277.

Pattnaik, S. K., Y. Sahoo and P. K. Chand. 1995. Efficient plant retrieval from alginate encapsulated vegetative buds of mature Mulberry trees. Sci. Hortic. 61: 227-239.

Paz Sanchez-Seco MP, Rosario D, Quiroz E, Guzman G, Tenorio A. A generic nested-RT-PCR followed by sequencing for detection and identification of members of the alphavirus genus. Journal of Virological Methods. 2001; 95(1-2): 153-161.

Philip T, Govindaiah, Bajpai AK, Datta RK, (1992) Chemical control of mulberry disease - a review. Indian J. Seric., 32: 42-45.

Philip T, Gupta VP, Govindaiah, Bajpai AK, Datta RK, (1994) Diseases of mulberry in India- Research priorities and management strategies. Int. J. Tropical Plant Diseases 12: 1-21

Philip T, Mala VR, Kamble CK, (2009) Role of botanicals in the management of mulberry pests and diseases- a review. Sericologia, 49(2): 151-160

Philip T, Sarkar A, Govindaiah, (1996) Screening of some promising genotypes of mulberry for leaf spot and rust resistance. Indian J. Seric. 35(2): 158-159.

Philip T, Govindaiah, Bajpai AK, Datta RK, (1992) Chemical control of mulberry disease - a review. Indian J. Seric., 32: 42-45.

Ponnuvel KM, Nakazawa H, Furukawa S, Asaoka A, Ishibashi J, Taaka H, Yamakawa M. A Lipase isolated from the silkworm shows antiviral activity against NPV. Journal of Virology. 2003; 77(19): 10725-10729.

Prasad KV, Dayakar Yadav BR, Sullia SB, (1993) Taxonomic status of rust on mulberry in India. Current Science, 65(5): 424-425.

Prasong S, Wilaiwan S, Nualchai K. Structure and Thermal Characteristics of Bombyx mori Silk Fibroin Films: Effect of different organic solvents. International Journal of Chemical Technology. 2003; 2: 21-27.

Pratheesh Kumar PM, Chattopadhyay S, Maji MD, Raje Urs S, (2005) Severity of foliar diseases of mulberry in West Bengal during commercial crop seasons. Bull. Ind. Acad. Seric. 9(1): 1-6.

Pratheesh Kumar PM, Maji MD, Gangwar SK, Das NK, Saratchandra B, (2000) Development of leaf rust (Peridiopsora mori) and dispersal of urediospores in mulberry (Morus spp.). International J. Pest Management. 46(3): 195-200.

Pratheesh Kumar PM, Pal SC, Qadri SMH, (2003). Screening of mulberry germplasm for resistance to leaf spot caused by Myrothecium roridum. J. Mycopathol. Res., 41(1): 105-106.

Pratheesh Kumar PM, Qadri SMH, Gangwar SK, Saratchandra B, (2000) Biological control of Phyllactinia corylea (Pers.) Karst, using composted plant extracts. Indian J. Seric., 39(1): 81-83.

Pratheesh Kumar PM, Qadri SMH, Pal SC, (2011) Factors influencing development and severity of grey leaf spot disease of mulberry (Morus spp.). International Journal of Industrial Entomology, 22(1): 11-15.

Pratheesh Kumar PM, Qadri SMH, Pal SC, Misra AK, (2011)Quantification of relation between disease intensities and physiological and biochemical changes in mulberry due to grey leaf spot. Indian J. Sericulture. 50(1): 28-33.

Pratheesh Kumar PM, Setua GC, (2004) Bacillus lentimorbus a potential biocontrol agent of collar rot pathogen (Sclerotium rolsfii) of mulberry. Bull. Acad. Seric., 7(1): 108-111.

Pratheesh Kumar PM, Qadri SMH, Pal SC, Saratchandra B, (1999) Evaluation of few fungicides against two leaf spot diseases of mulberry (Morus spp.). Bull. Sericult. Res., 10: 9-15.

Pratheesh Kumar PM, Chattopadhyay S, Maji MD, Raje Urs S, (2005) Severity of foliar diseases of mulberry in West Bengal during commercial crop seasons. Bull. Ind. Acad. Seric. 9(1): 1-6.

Pratheesh Kumar PM, Qadri SMH, Gangwar SK, Saratchandra B, (1999) Enhanced efficacy of fungicides using sticker in foliar mulberry disease control. Sericologia, 263-266.

Pratheesh Kumar PM, Qadri SMH, Pal SC, Mishra AK, and Raje Urs S, (2003) Post infection physiobiochemical alteration at various intensities of leaf spot (Myrothecium roridum) in mulberry. International J. Indust. Entomol., 7(2): 175-180.

Pratheesh Kumar PM, Maji MD, Qadri SMH, Saratchandra B, (2009) Bacterial biocontrol agents as alternate fungicides for management of two foliar diseases of mulberry. Proceedings of the International Conference on Emerging Technologies in Environmental Science and Engineering. Oct. 26-28, Aligarh Muslim University, Aligarh, India. pp 984-991.

Qadri SMH, Pratheesh Kumar PM, Gangwar SK, Elangovan C, Maji MD, Saratchandra B, (1998) Crop loss assessment due to powdery mildew in mulberry. Bull. Seric. Res. 9: 31-35.

Raja Gopal Reddy C, Nirmala RS, Ramanamma CH, (2009)Efficacy of phytoextracts and oils of certain medicinal plants against Cercospora moricola Cooke., inciting mulberry (Morusalba L.) leaf spot. Journal of Biopesticides, 2(1): 77-83.

Rajaduari, S. and Thiagarajan, V. (2003) Mulberry sap sucking insects. *Indian Silk,* 8: 5-8.

Rajkhowa G, Chakravorty R (2004). Studies on the intensity and seasonal population variation of whitefly, Dialeuropora decempuncta (Quaintance & Baker) on mulberry, Morus spp. in relation to abiotic factors. Indian J. Seric., 43 (2): 204-206.

Rangaswami, G. Narasimhanna, M.N., Kasi Viswanathan K. & Sastry, C.R. 1976. Manual on Sericulture Vol.1, Rome, FAO.

Ravikumar, J., Sakthivel, N. and Balakrishna, R. (2010) Incidence of black scale insect in mulberry. *Indian Silk,* 1(1): 4-5.

Ravindran, S., Ananda Rao, A., Girish Naik, V., Tikader, A., Mukherjee, P. & Thangavelu, K. 1997. Distribution and variation in mulberry germplasm. Indian J. Plant Genet. Res. 10(2): 233-242.

Ravindran, S., Ananda Rao, A., Girish Naik, V., Tikander, A., Mukherjee, P. & Thangavelu, K. 1997. Distribution and variation in mulberry germplasm. Indian J. Plant Genetic Resources, 10(2): 233-242.

Ravindran, S., Tikader, A., Girish Naik, V., Ananda Rao, A. & Mukherjee, P. 1988. Distribution of mulberry species in India and its utilisation. Poster paper presented in National Dialogue.

Reddy CR, Reddy PL, Misra S, Reddy KD, Sujathamma P, Reddy PR (2003) Sooty mould infection on mulberry –management. Intrl. J Industrial Entomology. 6(2): 203-205.

Reddy RM, Prasad GV. Silk- The prospective and compatible bio-material for advanced functional applications. Trends in Applied Sciences Research. 2011;6(1): 89-95.

Reddy RM. Innovative and multidirectional applications of natural fibre, silk – A Review. Academic Journal ofEntomology. 2009;2(2): 71-75.

Rubinstein R, Harley EH, Losman M, Lutton D. The nucleic acids of viruses infecting *Heliothis armigera*. Virology. 1976; 69: 323-326.

Sajeevan, R. S., Karaba, N.N., Shivashankara, K.S., Pallavi,N., Gurumurthy, D.S. and. Shivanna,M. B.2017. Expression of Arabidopsis *SHN1* in Indian Mulberry (*Morusindica* L.) Increases Leaf Surface Wax Content and Reduces Post-harvest Water Loss. *Front. Plant Sci.* 8: 418.

Sakthivel, N. and Qadri, S.M.H. (2010) Efcacy of certain insecticides and botanicals against mulberry thrips, *Pseudodendrothrips mori* Niwa (Thysanoptera: Thripidae). *Indian Journal of Entomology,* 72(2): 152-154.

Sakthivel, N. and Qadri, S.M.H. (2010) Incidence of papaya mealybug on tapioca in Tamil Nadu, *Indian Silk,* 1(5-6): 8-9.

Sakthivel, N., Mukund V. Kirsur and Balakrishna, R. (2010) Predatory fauna of papaya mealybug, *Paracoccus marginatus* Williams and Granara de Willink infesting mulberry. *Insect Environment,* 16 (3): 115-117.

Sakthivel, N., Mukund V. Kirsur, Punithavathy, G. and Balakrishna, R. (2010) Papaya mealybug, *Paracoccus marginatus* Williams and Granara de Willink (Hemiptera: Pseudococcidae) on mulberry in Tamil Nadu. *Insect Environment,* 16 (3): 117-118.

Sakthivel, N., Punithavathy, G. and S.M.H. Qadri (2009) Record of natural enemy Complex of spiralling whitey, *Aleurodicus dispersus* Russell (Homoptera: Aleyrodidae) infesting mulberry in Tamil Nadu. *Insect Environment,* 15(2): 80-82.

Samson MV, Baig M, Sharma SD, Balavenkatasubbaian M, Sasidharan TO, Jolly MS. Survey on the relative incidence of silkworm diseases in Karnataka, India. Indian Journal of Sericulture. 1990; 29(2): 248-254.

Samson MV, Nataraju B, Baig M, Krishnaswamy S. Starvation of *Bombyx mori* L., on cocoon crop and incidence of loss due to diseases. Indian Journal of Sericulture. 1981; 20: 42-47.

Santha Kumar MV, Bandyopadhyay UK (2001-02). Biological control of whitefly. Annual Report, Central Sericultural Research & Training Institute, Berhampore, West Bengal, India, Pp. 46

Sarkar DD. The silkworm biology genetics and breeding, Vikas Publishing House Pvt. Ltd. Jangpora, New Delhi. 1998, 1-338.

Sarkar, A. & Jalaja S. Kumar, 1996. Gradual improvement of mulberry varieties under irrigated conditions in South India and potential programme for varietal selection. International Symposium on Sericulture Science, Hangzhou, China.

Sarkar, A., Jalaja S. Kumar & Datta, R.K. 1996. Gradual improvement of mulberry varieties under irrigated condition in South India and the optimal programme for varietal selection in the tropics. Sericologia (in press).

Sarkar, A., Jalaja S. Kumar & Datta, R.K., 1999. Potentiality of Victory-1 under irrigated conditions of South India. Indian Silk, May, 1999.

Sastry, C.R. 1984. Mulberry varieties, exploitation and pathology. Sericologia, 24(3): 333-359.

Sato M, (1983) Search for pant leaves containing phage inducers effective to Pseudomonas syringae pv. mori, the causal agent of bacterial blight of mulberry. J. Seric. Sci. Japan, 52(4): 361-362.

Sato M, Takahashi K, (1973) Ecological studies on the bacterial blight of mulberry 2. Seasonal differences of the population of Pseudomonas mori phage in the soil of mulberry field. J Seric. Sci. Japan, 42 (3) : 207-212.

Sato M, Takahashi K, (1973) Ecological studies on the bacterial blight of mulberry 3. Some factors concerning with the survival of Pseudomonas mori (Boyer et Lambert) Stevens in soil J. Sericult. Sci. Japan, 42 (3): 213-218.

Sato M, Takahashi K, (1975) Ecological studies on the bacterial blight of mulberry I. The over wintering of the pathogen Pseudomonas mori (Boyer et Lambert) Stevens. J. Sericult. Sci. Japan, 41(4): 285-293.

Sato, M, Takahashi K, (1975) Ecological studies on the bacterial blight of mulberry VI. Duration of survival of Pseudomonas mori (Boyer et Lambert) Stevens and its phage in mulberry tissues and soil. J. Sericult. Sci. Japan, 44(2): 98-104.

Savanurmath CJ, Basavarajappa S, Hinchigeri SB, Ingalhalli SS, Singh KK, Sanakal RD. Relative incidence of the silkworm viral diseases in agroclimatic zones of northern Karnataka, India. Bulletin of Sericultural Research. 1994; 5: 51-55.

Schmidt, R., Oconnor-Sanchez, A., Willmitzer, L. and Rocha-Sosa, M. 1990. Construction of an intron- containing marker gene: splicing of the intron in transgenic plants and to use in monitoring early events in Agrobacterium-mediated plant transformation. *Mol Gen Genet.,* 8(20): 245–250.

Sen K, Babu KM. Studies on Indian silk I. macro characterization and analysis of amino acid composition. Journal of Applied polymer science. 2004;92: 1080-1097.

Sen K, Babu KM. Studies on Indian silk II. Structure- property correlations. Journal of Applied polymer science. 2004;92: 1098-1115.

Sengupta K. Diseases and pests, a scourge of tropical sericulture. Indian Silk. 1988; 26(10): 11-12.

Sengupta, K., Govindaiah and Kumar. P., (1991) Diseases and Pests of Mulberry and their Control. Central Sericultural Research and Training Institute, Srirampura, Mysore - 570 008.

Sengupta, K., Kumar, P., Baig, M. and Govindaiah (1990) Handbook on Pest and Disease Control of Mulberry and Silkworm. United Nations Economic and social commission for Asia and the Pacic, Bangkok, Thailand. Pp. 17-40.

Seni, A. and Sahoo, A.K. (2014) Biology of the papaya mealybug, *Paracoccus marginatus* Williams and Granara de Willink (Hemiptera: Pseudococcidae) *International Journal of Agriculture, Environment & Biotechnology,*7(4): 875-881.

Shabir Ahmad Bhat., IFAT Bashir and AFIFA S. Kamili., 2013, Microsporidiosis of silkworm, Bombyx mori L, Afri, J of Agricultural Research Vol. 4 (13), pp. 1519-1523.

Shamim M, Baig M, Nataraju B, Datta RK, Gupta SK. Evaluation of protein-a linked monoclonal antibody latex agglutination test for diagnosis of nuclear polyhedrosis virus (*Bm*NPV) of silkworm *Bombyx mori* L. Journal of Immunoassay and Immunochemistry. 1995; 16(2): 155- 66.

Shapiro A, Green T, Rao S, White S, Carner G, Mertens PP, Becnel JJ. Morphological and molecular characterization of a cypovirus (Reoviridae) from the mosquito *Uranotaenia sapphirina* (Diptera: Culicidae). Journal of virology. 2005; 79(15): 9430-9438.

Sharma D.D, Govindaiah, (1991) Comparative in vitro toxicity of fungicides against Pseudomonas mori (Boyer & Lambert) Stevens - causing bacterial blight of mulberry. Indian J. Seric., 30(2): 139-140

Sharma DD, Chowdary NB, Nishttha Naik V, Mala VR, Dandin SB, (2005) Blight diseases of mulberry (Morus spp.)-a review. Sericologia, 45(3): 231-245.

Sharma DD, Govindaiah, (1991). Comparative in vitro toxicity of fungicides against Pseudomonas mori (Boyer and Lambert) Stevens causing bacterial blight of mulberry. Indian J Seric. 30 (2): 139-140.

Sharma DD, Gupta VP, Baqual MF, Chandrashekar DS, (2001) Efficacy of plant extracts against Pseudomonas syringae pv. mori and Xanthomonas campestris pv. mori causing bacterial blight in mulberry. Indian J. Seric 40(1); 109-111.

Sharma DD, Mala VR, Das PK, Dandin SB, (2008) Effect of composts enriched with useful/antagonistic microbes on suppression of nursery diseases in mulberry. Sericologia, 48 (1): 101-108.

Sharma DD, Nishitha Naik V, Chowdary NB Mala VR, Kamble CK, (2009) management of mulberry diseases through eco-friendly approaches- a review. Sericologia, 49(2): 123-138.

Sharma DD, Nishitha Naik V, Chowdary NB, Mala VR, (2003) Soilborne diseases of mulberry and their management - a review. Int. J. Indust. Entomol., 7(4): 93-106.

Sharma PD, (1974) A new leaf spot disease of mulberry in India. Curr. Sci., 43: 429.

Sheikh MRK, Farouqui FI, Modak PR, Hoque MA, Yasmin Z. Dyeing of raishahi silk with basic dyes: effect of modification on dyeing properties. Journal of the Textile Institute. 2006;4(97): 295-305.

Sherald JL, Hearon SS, Kostka SJ, Morgan DL (1983) Sycamore leaf scorch: Culture and pathogenicity of fastidious xylem-limited bacteria from scorch-affected trees. Plant Disease, 67: 849-852

Sherald JL. (Plant Pathologist, Center on Urban Ecology, National Park Service, Washington, D.C. (1988). New Bacterium Identified as Cause of Scorch Symptoms. American Nurseryman 167(10): 15, May 15, 1988.

Shery, A.V.M.J. (2014) The predominant thrips species infesting mulberry in Karnataka, India. *Insect Environment,* 20(3): 88-92.

Shery, A.V.M.J. and Bindroo, B.B. (2013) Oberea artocarpi Gardner (Cerambycidae - Coleoptera) an emerging pest on mulberry in Kerala and Karnataka. *Insect Environment,* 19(3): 191-192.

Shimura K. The structure, synthesis and secretion of fibroin in silkworm, Bombyx mori, Sericologia. 1988;28(4): 457-479.

Shoukang, L., J. Dongfeng and Q. Jun. 1987. In vitro production of haploid plants from mulberry (Morus) anther culture. Scientia. Sinica. 30: 853-863.

Siddaramaiah A L, (1987) A new root rot, bud and leaf blight of mulberry seedlings from India. Ind. J. Seric., 27(2): 161.

Siddaramaiah AL, Govindan R, Desai SA, Bhat RP, Devaiah MC, (1980) In vitro efficacy of certain systemic and non systemic fungicides against Cercospora leaf spot of mulberry Ind. J. Seric. 19: 34-35.

Siddaramaiah AL, Krishna Prasad KS, (1978) Laboratory evaluation of fungicides against Cercospora moricola Cooke. Ind. J. Seric., 17: 33-36.

Siddaramaiah AL, Lingaraju S, Hegde RK, (1978) Toxic effect of the culture filtrate of Cercospora moricola Cooke the causal organism of leaf spot of mulberry. Ind. J. Seric., 19: 32-33.

Siddaramaiah, AL, Srikanth Kulkarni, Krishna Prasad KS, (1978) Antagonistic action of Streptomyces sp. on Cercospora moricola Cooke. Ind. J. Seric., 17 (1): 64.

Sikdar AK, Krishnaswami S, (1980) Assessment of leaf yield loss of the two mulberry varieties due to leaf spot disease. Ind. J. Seric. 19: 9-12.

Sikdar AK, Shenoi MM, (1980) Control of leaf spot disease of mulberry by systemic fungicides. Indian Phytopath., 33(1): 38-41.

Sing, R.N., Santha, P.C., Sasidharan, T.O., Manjula, A. and Kamble, C.K. 2004. Delayed mother moth testing for effective detection of pebrine. Indian Silk, 43: 7-8.

Singh KVP. Introduction to Textiles, 1st Ed. New Delhi, Kalyani Publishers. 2004, 131p.

Singh T, Saratchandra B (2003). Microsporidian disease of the silkworm Bombyx mori L. (Lepidoptera: Bombycidae) Int. J. Indus. Entomol. 6: 1-9.

Singh, R.N. and Saratchandra, B. (2011) Sericultural Entomology. A.P.H. Publishing Corporation, New Delhi.

Singh, R.N., Maheshwari, M. and Saratchandra, B. (2005) Biology and control of whiteies in sericulture. *Insect Science,* 12(6): 401 – 412.

Singh, R.N., Samson, M.V. and Datta, R.K. (2000) Pest Management in Sericulture. Indian Publishers Distribution, Delhi.

Singh, T. and Singh, P.K. (2013) Mulberry Crop Protection. Discovery Publishing House (DPH) Pvt. Ltd., New Delhi.

Sonwalkar TN. Handbook of Silk Technology, New Age International Limited, Publishers, 1993, 10p.

Sprague V (1982). Microspora. In "Synopsis and Classification of Livingorganisms" (S.P Parker, ed.) Vol. 1. pp. 589-594, McGraw-Hill, New York

Srikanta, H.K., 1986, Studies on cross infectivity and viability of Nosemabombycis Naegeli (Microsporidia: Nosematidae), M.Sc. (Agri.) Thesis, UAS, Banga1ore, p.106.

Srivastav PK, Thangavelu K. Sericulture and Seri- biodiversity. New Delhi, Associated Publishing Company, 2005, 1-254.

Stasolla, C. and E. C. Yeung. 2003. Recent advances in conifer somatic embryogenesis: improving somatic embryo quality. Plant Cell Tiss. Org. Cult. 74: 15-35.

Steinhaus EA. Stress as factor in insect diseases. Proc. Tenth International Congress Entomology. 1958; 4: 725- 730.

Strydom DJ, Haylet T, Stead RH. The amino terminal sequence of silk fibroin peptide ep. A reinvestigation. Biochem Bio-Phys. Res. Commun. 1977;3: 932-938.

Sudhakar R, Ninge GKN. A natural colourant from ficus racemose for silk. Colourage. 2011;58(9): 69.

Sugimori H, Nagamane T, Kobayashi M. Analysis of structural polypeptides of *Bombyx mori* L. (Lepidoptera: Bombycidae), nuclear polyhedrosis virus. Applied Entomology Zoology. 1990; 25(1): 67-77.

Sukumar J, (1988) Inhibition of conidial germination of Cercospora moricola by mulberry leaf leachates. Sericologia, 28(2): 249-252.

Sukumar J, Ramalingam A, (1986) Epidemiology of Cercospora leaf spot disease of mulberry stages of infection spots and conidial production. Sericologia 27(4): 685-688.

Sukurnar J, Padma SD, Bongale VD, (2003) Occurrence of streptomycin resistant pathogenic strains of Pseudomonas mori in mulberry gardens of Karnataka. Agric. Sci. Digest., 23 (2) : 140 – 142.

Sundareswaran P, Govindaiah, Srinivasan EB, Jolly MS, (1988) Effect of leaf rust disease on the nutritive composition of mulberry (Morus alba L.). Indian J. Seric. 27(2): 159-160.

Sunil Kumar T., Narayanaswamy T. K and Asha I.S (2018) 'Gene Transfer Technique in Mulberry Crop Breeding', *International Journal of Current Innovation Research,* 4(6), pp. 1261-1264.

Sunil Kumar, T., Narayanaswamy, T.K. and Manjunatha, H.A. (2018) Mulberry defoliating insect pests and their management. *International Journal of Current Innovation Research,* 4(5): 1228-1231.

Susheelamma, B.N., Jolly, M.S., Sengupta, K., Giridhar, K., Baksh, S., Mogili, T., Mallikarjuna, R.S. & Pavan Kumar, T. 1992. Statistical analysis of adaptability of drought resistant mulberry genotypes. Sericologia, 32(4): 619-628.

Suzuki T, Kitano M. Lateral bud development and shoot growth in low pruned Morus alba as affected by stem orientation. J. Seric. Sci. Japan. 1989;58: 1- 4.

Takahashi K, (1975). Chemical control of mulberry diseases in Japan. Japan Pesticide Information, 25: 27-31.

Takahashi K, (1980) The causal pathogen of bacterial blight of mulberry and its control. JARQ, 14(1): 41-45.

Takahashi K, (1995) Mulberry diseases and its control. Japanese Pesticide Information, 25: 1-6.

Talukdar B, Saikia M, Handique PJ, Devi D. Effect of organic solvents on tensile strength of muga silk produced by Antheraea assamensis. International Journal of Pure and Applied Sciences and Technology. 2011;7(1): 81-86.

Tanada, T., 1963, Epizootiology of Infectious Diseases. In: Insect Pathology-An Advanced Treatise (Ed: E.A. Steinhaus). Vol II, Academic Press, New York, pp.423- 475.

Tang AK, Salam KA, Samad MA, Absar N, (2005) Nutritional changes of four varieties of mulberry leaves infected with fungus (Cercospora moricola). Pakistan J. Biological Sciences. 8(1): 127-131.

Tatsuke K (1971). Microsporidiosis disease of Silkworm a technical report, Overseas Technical Cooperation agency, Tokyo, Japan, pp.1-2.

Tayal MK, Chauhan TPS. Silkworm diseases and pests. Industrial Entomology, 2017, 265-289.

Teotia RS, Kant AK, Mandal SK, (1992) Bavistin-an effective fungicide for sericulture. Indian Silk, May 1992, pp 25-26.

Teotia RS, Mandal SK, (1991) controlling the leaf rust disease of mulberry, Indian Silk, April 1991, pp12-14

Teotia RS, Mandal SK, (1993) Bacterial leaf blight disease of mulberry. Indian Silk, October, 1993. pp 42-44.

Teotia RS, Rao KM, Mandal SK, (1994) Influence of growth regulators on the development of leaf rust and powdery mildew disease of mulberry. Sericologia, 33(4): 669- 673.

Teotia RS, Sen SK (1994) Mulberry diseases in India and their control. Sericologia, 34(1)1-8.

Teotia RS, Sen SK, Mandal SK. Sinha SS, (1994) Varietal resistance in mulberry to leaf rust (Aecidium mori Berclay). Sericologia, 34(4): 759-763.

Teotia RS, Mandal SK, (1992). Integrated control of leaf spot disease of mulberry. Indian silk, January 1992 pp 45-47

Thomas, T. D., A. K. Bhatnagar and S. S. Bhojwani. 2000. Production of triploid plants of mulberry (Morus alba L) by endosperm culture. Plant Cell Rep. 19: 395-399.

Thomson, H.M., 1960, J. Invertebr. Pathoi., 2: 346-385.

Tipton, J. 1994. Relative drought resistance among selected southwestern landscape plants. J. Arboriculture, 20(3): 151-155.

Tsukada M, Freddi G, Minoura N, Allara G. Preparation and application of porous silk fibroin materials, Journal of Applied Polymer Science. 1996;54: 507-514.

Tsukada M, Freddi G, Nagura M, Ishikawa H, Kasai N. Structural- changes of silk fibres induced by heat treatment. Journal of applied polymer science. 1992;46(11): 1945-1953.

Tyagi, K. and Kumar, V. (2011) A new record of pest species *Pseudodendrothrips bhattii* Kudo (Thysanoptera; Thripidae; Dendrothripinae) from India. *Indian Journal of Entomology.* 73(4): 296-297.

Ude AU, Eshkoor RA, Zulkifili R, Ariffin AK, Dzuraidah AW, Azhari CH. Bombyx mori silk fibre and its composite: a review of contemporary developments. Materials and Design. 2014;57: 298-305.

Ullal, S.R. & Narasimhanna, M.N. 1987. Handbook of Practical Sericulture, Bangalore, Central Silk Board.

Umesh. Kumar NN, Shree MP, (1990) Some physiological changes in leaves of Morus alba L. under pathogenesis. Geobios, 17: 140-141.

Undeen, A.H. 1997. A handbook of biology and research technique. In: Microsporidia (Protozoa) Bulletin No.387 (Ed. J.D. Millar), Southern Cooperative Series, Oklahama University, USA.

V Sivaprasad, L Satish, G Mallikarjuna, N Chandrakanth, AV Mary Josepha, SM Moorthy (2021). A Field-Friendly Loop-Mediated Isothermal Amplification (FF-LAMP) method for rapid detection of Nosema bombycis in silkworm, Bombyx mori. ISJ 18: 66-74, 2021

Vago C, Atger P. Multiplication massive des virus'D insectes pendant la mue numphale. Entomophaga. 1961; 6(1): 53-56.

Vanapruk P, Attathom T, Sanbatsiri K, Attathom S. Comparison of methods for the detection of nuclear polyhedrosis virus in silkworm, *Bombyx mori* Linn. In: Proceedings of the 30th Kasetsart University Annual Conference: Plant Science, Kasetsart University, Bangkok, Thailand, 1992, 237-243.

Vanregenmortel MHV, Fauquet DH, Bishop L, Carstens EB, Estes MK, Lemon SM, Maniloff J, Mayo MA, McGeoch DJ, Pringle CR, Wickner RB. Virus Taxonomy: Classification and Nomenclature of Viruses, 7th Report of the International Committee on Taxonomy of Viruses, Academic Press, San Diego, California, 2000, 309-343.

Venugopala Pillai S, Krishnaswamy S. Effect of high temperature on the survival rate, cocoon quality and fecundity of *Bombyx mori* L. Proceedings of Sericultural Symposium and Seminar, TNAU, Coimbatore, India, 1980, 141-148.

Vijaya Kumar GC, (1994) Incidence of Cercospora moricola Cooke in mulberry leaves (Morus Indica L.): Histopathological and histochemical changes. Sericologia, 34(2) 363-366.

Vijaya Kumari, N. (2014) Eco-friendly technologies for disease and pest management in mulberry-A Review. *Journal of Agriculture and Veterinary Science.* 7(2): 01-06.

Vijayakumar GC, (1991) Histopathological and histochemical changes in mulberry (Morus alba L.) powdery mildew due to infection by Phyllactinia corylea (Pers.) Karst. Sericologia, 31(4): 703-707.

Vijayan, K. and S. N. Chatterjee. 2003. ISSR profiling of Indian cultivars of mulberry (Mores spp.) and its relevance to breeding programs. Euphytica..131: 53-63.

Vineet Kumar, Babu AM, Sharma DD, Datta RK, (1998) Penetration and infection process of Phyllactinia corylea on mulberry leaf causing powdery mildew: I- A sexual stage. J. Phytopath., 146: 469-472.

Vineet Kumar, Dhar A, Gupta VP, Babu AM, Sarkar A, Datta RK, (2001) SEM observations on the perithecia of Phyllactinia corylea causing powdery mildew disease in mulberry. Int. J. Indust. Entomol., 3(1): 63-67.

Vineet Kumar, Gupta VP, (2004) Scanning Electron Microscopy on the perethecial development of Phyllactinia corylea on Mulberry-II. Sexual stage. J Phytopathology, 152: 169-173.

Vineet Kumar, Katiyar RL, Babu AM, Sarkar A, Datta RK, (2000). Light and SEM studies on the life stages of Illeis Indica (Timb.) and its predation on Phyllactinia corylea (Pers.) Karst. Infecting mulberry. Arch. Phytopath. Pflanz., 33: 149-159.

Wafa, A.K. And Kotby, F.A., 1971, Bull. Em. Soc. Egypt., 55: 259-274.

Wan WB, Fei JM, Wu Y, Bai XC, Yu F, Shi GF, Li YF, Kuai YZ, (2010) a new report of a mosaic dwarf viroid like disease on mulberry trees in China. Polish J. Microbiology, 59(1): 33-36.

Watanabe H, Maeda S, Matsui M, Shimizu T. Histopathology of the midgut epithelium of the silkworm, *Bombyx mori*, infected with a newly-isolated virus from the flacherie-diseased larvae. The Journal of Sericultural Science of Japan. 1976; 28;45(1): 29-34.

Watanabe H, Maeda S. Multiplication of a densonucleosis virus in the silkworm, *Bombyx mori* L., reared at high temperature. Japanese journal of applied entomology and zoology, 1979, 151-155.

Watt, G. 1981. Dictionary of economic products of India Vol. 5, p. 279-284. Delhi, Periodical Experts.

Wells JM. (USDA, ARS, Plant Pathologist, Rutgers University, New Brunswick, NJ 08903), Boligala C. Raju, Hsueh-Yun Hung, William G. Weisburg, Linda Mandelco-Paul, and Don J. Brenner. 1987. *Xylella fastidiosa* gen. nov., sp. nov: Gram-Negative, Xylem-Limited, Fastidious Plant Bacteria Related to *Xanthomonas spp*. International Journal of Systematic Bacteriology, 37 (2) 136-143.

Yamanouchi, H., A. Koyama and H. Machii. 1999. Effect of medium conditions on adventitious bud formation in immature mulberry leaves. J.A.R.Q. 33: 267-274.

Yang, J. H. and X. H. Yang. 1989. Breeding of artificial triploids in mulberry. *Seric. Sci. Jpn.* 15: 65-70.

Yoshida Z, Kato M. Journal of Chemical Society of Japan. 1955;58(274): 667.

Index